Research, Evaluation and Audit

Research, Evaluation and Audit

Editor

Roger Burnett

scitus
academics

Research, Evaluation and Audit
Edited by **Roger Burnett**

ISBN: 978-1-68117-638-3
Library of Congress Control Number: 2016946498

616, Corporate Way, Suite 2, 4766,
Valley Cottage, NY 10989

www.scitusacademics.com

Notice

Printed in United States of America on Acid Free Paper

Preface

Research and evaluation are characterized by similar features that center on the shared objective of answering a question. Although both use similar data collection and analysis methods, the two disciplines diverge again during use and dissemination. Research is intended to increase the body of knowledge on a particular issue; any subjective opinion limits the researcher's credibility. On the other hand, evaluators must balance the need to remain objective and the expectation to make recommendations for stakeholders. Evaluators must determine what information is valuable, what method is best for data collection, how to analyze the data, and how to relay findings to stakeholders. This requires interpretation and a certain level of judgment by the evaluator that is absent from the role of the traditional researcher. Evaluating the authority, usefulness, and reliability of the information you find is a crucial step in the process of library research. You will need to evaluate each resource you use for research, whether it is an online or print journal article, a website, a book, a newspaper article, or other source that you want to cite.

Research, Evaluation and Audit considers methods and approaches that are suitable in a library context, making it a useful tool for library professionals and practitioners alike. This book is a thorough introduction to research and evaluation in libraries. A large part of the book is dedicated to a variety of research methods, including historical, descriptive, and experimental methods. Whether student or practitioner, information professionals in all types of libraries can benefit from incorporating research and evaluation into their regular planning activities.

TABLE OF CONTENTS

Chapter 1 Research Data Management Services in Academic Research Libraries and Perceptions of Librarians 1

Chapter 2 Where are We going? Overview of Major Collaborative Conservation Research Projects and Implications for the Future Preservation of Collections 29

Chapter 3 Preparing for New and Changing Roles in Research Libraries – the Need for Continuing Professional Development 43

Chapter 4 Designing Libraries for Research Collaboration in the Network World: An Exploratory Study 63

Chapter 5 On the Role of Research Data Centres in the Management of Publication-Related Research Data 105

Chapter 6 Library Roles in University Research Assessment 133

Chapter 7 Evaluation as a Tool for Developing the Quality of the Academic Libraries. Case Study at a Large and Traditional Research Oriented Scandinavian

University - Viewpoint of a Member of the Evaluation Panel 155

Chapter 8 Information Society and Library Evaluation Transitions in Portugal: A Meta-evaluation Model and Frameworks (1970–2013) 161

Chapter 9 A Global Approach to Digital Library Evaluation 189

Chapter 10 "Just How Are We Going to Satisfy Our Research Customers" 203

Chapter 11 Information Handling Iin Collaborative Research 211

Chapter 12 Supporting Teaching and Research in an Online Environment: Developing the University of London Library Model 231

Chapter 13 Evaluation as a Tool for Developing the Quality of Academic Libraries. Case Study at a Large and Traditional Research Oriented Scandinavian University 249

Index 267

1

CHAPTER

Research Data Management Services in Academic Research Libraries and Perceptions of Librarians

Carol Tenopir[a], Robert J. Sandusky[b], Suzie Allard[a], Ben Birch[a],

[a] School of Information Sciences, University of Tennessee, 451 Communications Bldg., 1345 Circle Park Drive, Knoxville, TN 37996-0341, USA

[b] Richard J. Daley Library, MC-234, 801 S. Morgan St., Chicago, IL 60607, USA

HIGHLIGHTS

- Research data management services (RDS) in academic libraries are explored, along with librarian perceptions.
- North American libraries do not frequently offer RDS, but many are being planned.
- Technical (hands-on) RDS are less common than informational (consulting) RDS.
- Over half of academic libraries do not collaborate with any other RDS provider.
- Many librarians in academic research libraries feel the need for more RDS training.

ABSTRACT

The emergence of data intensive science and the establishment of data management mandates have motivated academic libraries to develop research data services (RDS) for their faculty and students. Here the results of two studies are reported: librarians' RDS practices in U.S. and Canadian academic research libraries, and the RDS-related library policies in those or similar libraries. Results show that RDS are currently not frequently employed in libraries, but many services are in the planning stages. Technical RDS are less common than informational RDS, RDS are performed more often for faculty than for students, and more library directors believe they offer opportunities for staff to develop RDS-related skills than the percentage of librarians who perceive such opportunities to be available. Librarians need opportunities to learn more about these services either on campus or through attendance at workshops and professional conferences.

1. INTRODUCTION

Science has entered a "fourth paradigm" that is more collaborative, more computational, and more data intensive (Hey, Tansley, & Tolle, 2009a) than the previous experimental, theoretical, and computational paradigms. This emerging scientific paradigm is often referred to as e-science or e-research (Hey, Tansley, & Tolle, 2009b). Increased reliance on technology in all parts of scientific endeavor, or cyberinfrastructure, and the establishment of data management and data sharing mandates by many research funding bodies[1] have motivated academic libraries to take action with regard to the shifting needs of their faculty and students and consider how best to engage in e-science through the development of library-based research data services (RDS). In the U.S. and Canada, individual large academic research libraries often lead these activities (Association of Research Libraries, 2010).

The results of investigations into (a) librarians' RDS practices in U.S. and Canadian academic research libraries and (b) the RDS-related library policies in the same type of libraries are reported here. These studies establish a baseline assessment of the RDS involvement of individual librarians as well as libraries as institutions. The results inform and enable practitioners, administrators, and educators to make strategic RDS plans in academic research libraries and guide the evolution of curricula in LIS education.

2. PROBLEM STATEMENT

The emerging need for research data management is prompting library directors to plan for additional RDS to be offered by their libraries, and at the same time many librarians are looking for opportunities to develop their RDS-related skills. But are library directors and librarians on the same page regarding RDS? In other words, do library policies in this regard align with librarians' perceptions? Misalignment can hinder effective start-up of RDS. This study focuses on the alignment issue by comparing data from library directors on RDS currently offered or planned, with data from librarians on RDS currently performed. Similarly, comparisons are made between library directors' perceptions on how their libraries are providing RDS development opportunities for staff and perceptions of librarians on the availability of such opportunities in their library. Insight on the part of the library community gained from this study could raise awareness of such misalignment, followed by corrective action leading to more efficient development of RDS.

This paper combines the findings from two surveys to answer several questions regarding North American academic research libraries and their involvement in RDS, including:

- How many academic libraries are actively engaged in RDS?
- How many academic libraries are planning to be involved in RDS in the near future?
- Are libraries offering opportunities for their librarians to gain RDS related skills?
- What specific types of RDS are being offered?
    - Which are more common — *informational, consulting-type services* (for example, helping faculty and students find a place to deposit their research data or pointing to data management plan examples) or *technical, hands-on services*(for example, running an institution-housed data repository or helping researchers write data management plans)?
    - Are libraries offering or thinking of offering a mix of services or concentrating on a single service?

Libraries may already be offering or planning to offer RDS and they may have developed plans to do so. However, it is the librarians who are on the front lines in terms of implementing these plans. Therefore, two separate studies were conducted; the first – the library study – surveyed directors of academic libraries in the U.S. and Canada and sought answers for the questions above. The second – the librarian study – surveyed a sample of librarians who work in academic research libraries. Librarians were surveyed in an attempt to answer such research questions as:

- How many academic librarians are currently involved with RDS in their libraries?
- Do librarians think they have the disciplinary background and training to offer RDS?
- Do librarians feel they have opportunities to learn what they need to know about RDS?

Finally, by studying both the directors of academic libraries to get official library policy and the front-line librarians who work at academic research libraries to get their perceptions and personal perspectives, the alignment between library practices and librarians' perceptions of the development opportunities open to them regarding research data management services can be evaluated. Although results from each of these individual studies have been previously reported, this article compares the results for the first time in order to determine if the practices and opportunities for training in research data service provision in academic research libraries are in alignment with the perceptions of the librarians as to their preparation and opportunities.

3. LITERATURE REVIEW

The literature related to this paper includes studies of *librarians, libraries,* and RDS; and papers that present case studies or recommendations for how librarians and libraries can develop RDS. In addition to surveys of the current status of RDS in libraries in several countries, current literature covers a range of recommended specific services. The literature shows that RDS or e-science services in libraries are discussed in the library literature, and are being offered by some research libraries, but are not yet being offered by most. Peters and Dryden (2011) found that the most important data services needed by researchers are mainly directional ones: grant proposal support including data management planning, locating data-related services, publication support, and specific data management assistance. Another study (Bach et al., 2012), however, found that, of the surveyed biodiversity data repositories, most only deliver low-level support for users.

Many librarians and researchers engaged with e-research have discussed the possible roles for both libraries and librarians in providing RDS (Association of Research Libraries, 2006, Council on Library and Information Resources, 2008, Gabridge, 2009, Gold, 2007,Hey and Hey, 2006 and Jones, 2009). One third of participants in a UK survey (Brown & Swan, 2007) believed that within five years "manager of datasets from e-science/grid projects" (p. 47) will be a major obligation of a librarian, with another third assigning it a secondary responsibility. MacColl (2010) advises libraries take on a more comprehensive and strategic role: libraries should be involved throughout the research process and need to be actively engaged in curating, advising, and preserving research outputs. Some additional suggested roles for libraries are to develop researchers' data-awareness, to adopt a data archiving and preservation role, and to train data librarians.

Libraries and librarians have a long way to go before realizing these roles. Potter, Cook, and Kyrillidou (2011) found that only 9 of 86 (or 10%) narrative profiles created by ARL members in the U.S. and Canada included references to "e-science/data curation and management" (p.7) as an important service supporting faculty success and scholarly communications. As part of a 2007 investigation into all of the types of support provided to researchers by 134 U.S. and Canadian academic health science libraries, Cheek and Bradigan (2010) found that just 12.2% of these libraries provided support for "data curation". About half of the respondents to ARL's 2009 North American e-science survey (Association of Research Libraries, 2010) had on-campus support units for scientific research data; however, the Data Working Group at Cornell University Library discovered that few university libraries were actually involved in research data curation (Steinhart et al., 2008).

A major aspect of data curation is preservation work. A 2009 survey of European data managers, of which nearly three-quarters (73%) were employed in libraries, found that the top three reasons for research data preservation included accountability for publicly funded research, inspiration for scientific advancements, and reanalysis of previously generated data (Kuipers & Van der Hoeven, 2009, p. 37). A majority of data managers, including those not employed by libraries, reported that their institutions have a policy for preservation of research data. These data managers did not report on the percentage of the data that was shared with other researchers. One Australian study found that data sharing is not a priority for many researchers (Markauskaite, Kennan, Richardson, Aditomo, & Hellmers, 2012). In this study, 864 researchers at seven Australian Universities were surveyed; 50% of participants did not allow access to any of their data and only 9% provided access to all of their data.

Research institutions have a responsibility to offer researchers educational and support services relating to data management and to encourage data sharing, in addition to providing policies and structure for research data preservation (Tenopir, Birch, & Allard, 2012). There is a need for more tailored and streamlined data services, but the identification of researchers' needs is difficult due to the complex nature of research data.Carlson (2012) at Purdue University Libraries, found that among and within fields of study there are disparities in the way that data curation is conceptualized and communicated. These variations make it challenging for librarians to understand the needs of researchers.

There are also many other challenges that librarians and libraries face in RDS development. Corrall, Keenan, and Afzal (2013) found clear evidence that development of specialized RDS is often constrained by knowledge and skills gaps among library staff and a lack of confidence in their expected roles in RDS. In a small-scale survey of New Zealand academic and college library managers, Brown (2010) found that there was little direct involvement in providing RDS but that libraries were participating in local steering groups, performing institutional planning, and involved in policy development both within and between academic institutions. This survey found that funding, librarian training, marketing, and uncertain demand from researchers and students were barriers to successfully providing RDS (Brown, 2010). Similarly, Creamer, Morales, Crespo, Kafel, and Martin (2012) found that health and science librarians have a high level of interest in developing a range of RDS skills, but often lack the skills needed to effectively provide RDS. More than half of the libraries were creating a "library strategic plan or policy for data management" (p.21), although they faced "serious barriers" to engaging in e-science, including funding for personnel and equipment and lack of broader institutional support, as well as "territorial struggles" between various other departments within the institution (p.23).

Recently, there have been tools developed and recommendations made to help overcome the challenges faced by librarians and libraries. The result of Carlson's (2012)terminology variation study was the development of the DCP Toolkit (http://datacurationprofiles.org). This tool enhances the data reference interview and enables librarians to connect with and discover the data needs of researchers. The detailed profiles in the DCP Toolkit provide insight into the data management language utilized by researchers in different fields. A number of studies have also acknowledged the importance of educating library staff about data curation and management services. Many library staff members have collection experience related to traditional materials, but may require training in relation to selecting and compiling data for inclusion in repositories. Research libraries have campus-wide faculty relationships and are proficient at developing conventional collections, giving them a competitive advantage in establishing a university's scientific data collection; however, Newton, Miller, and Bracke (2010) found that additional training is needed to build up an institutional data repository. Libraries need to utilize their professional connections with campus faculty, as well as faculty and staff at other institutions to collaborate and develop more skills in identifying appropriate materials.

4. METHODOLOGY

In order to compare like-to-like (that is, librarians in research libraries with policies of research libraries), the library data reported here are a subset of a larger study that surveyed all types and sizes of academic libraries in the U.S. and Canada. The full study sought to answer, in addition to the research questions addressed here, what sizes and types of academic libraries are most involved in RDS and how involvement varies by type and size of academic library. The full results were reported in Tenopir et al., 2012.

In the full study, survey responses were received from 223 library directors. In order to investigate the effects of variances in sizes and types of libraries, four demographic characteristics of the parent institution were used: number of full-time equivalent (FTE) students (less than 5000 vs. 5000 or more), number of tenure-track and tenured faculty (less than 100 vs. 100 or more), number of National Science Foundation (NSF) grants typically awarded per year (none vs. some), and type of institution (research or doctoral vs. baccalaureate vs. associate's).

Not surprisingly, academic research libraries, larger schools, and those receiving more NSF grants were more likely to be offering or planning to offer research data management services than other types of academic libraries (Tenopir et al., 2012).

Comparing results by these demographics also uncovered differences in methods libraries were using to develop staff capacity for RDS. Libraries at institutions with high enrollments, those with a large faculty, and those at research institutions were all more likely to have already reassigned or to be planning to reassign existing staff than libraries at other institutions.

Finally, there were considerable differences in library engagement with RDS. Libraries at institutions with high enrollment, larger faculty size, and at research institutions were more likely than libraries at smaller schools to be involved in things such as managing RDS technology infrastructure, planning RDS skills development opportunities for staff, and collaborating with other units on campus (Tenopir et al., 2012).

Because of these differences, the authors decided to examine in more depth only the results from libraries in research or PhD-granting institutions and to compare the results with another survey of individual academic librarians who work in research universities.

As in the academic library study, the librarian data reported here also represent a subset of a larger study. In the larger study, librarians employed by ARL member libraries were surveyed if their area of responsibility seemed likely to currently, or in the future, include RDS. The full study sought to answer how librarians' opinions of their preparedness to provide RDS, their library's support for their professional RDS development, the importance of RDS for libraries and their associated institutions, and the contributing or inhibiting factors for librarian involvement in RDS, varied with their current degree of engagement with RDS. The results of this part of the full study were reported in Tenopir, Sandusky, Allard, and Birch (2013).

The full results indicated consensus that the absence of RDS would adversely affect the institution's perception of the library in terms of relevance and prestige, that provision of RDS would augment the institution's research impact, and that the absence of RDS would put the institution at a disadvantage for grants. In addition, participants strongly rejected the idea that RDS would be a distraction and the idea that RDS are unnecessary and strongly affirmed that RDS fits the traditional role of librarians as stewards of scholarship (Tenopir et al., 2013).

The current analysis compares, for the first time, the frequency of RDS provision in academic research libraries with the services offered by the librarians.

The full library study was distributed to a stratified random sample of 351 library directors who are members of a panel organized by the Association of College and Research Libraries (ACRL). Each of these directors had agreed to participate in several ACRL surveys on assorted topics over the course of a year.

A total of 221 of these ACRL directors responded, for a response rate of 63%. Surveys were initially distributed in November 2011, with a follow-up in January 2012 (Table 1).

Table 1. Distribution to ACRL panel members.

Classification	Panel members	Responses	Response rate
Associate-degree granting	116	68	59%
Baccalaureate-degree granting	93	54	58%
Doctorate-granting	142	99	70%
Totals	351	221	63%

A separate distribution to several libraries in the University of California (UC) system yielded two additional responses. The number of invitations sent in the UC system, and therefore the exact response rate is unknown, although there are ten campuses in the UC system. The final dataset contains responses from 223 academic libraries. This paper focuses on the 101 responses from research/doctorate granting institutions (the 99 from ACRL distribution and the two UC universities; see Tenopir et al. (2012) for a full analysis of all 223 library responses).

In order to compare official library policy with the perspective of the librarians who work in academic research libraries, a separate survey was sent to a sample of academic librarians in the U.S. and Canada. The survey was sent to librarians who work in the 115 academic libraries that belong to the ARL. Between April 2011 and August 2011, a total of 948 invitations were sent to a sample of ARL librarians who work as subject librarians, metadata librarians, e-science librarians, or data librarians. This survey had 222 responses: a response rate of 23%.

In November 2011 and February 2012 a separate invitation was sent to librarians working in two libraries in the UC system, and to librarians whose ACRL library director volunteered to distribute the survey to their staff. This yielded 80 additional responses. The exact number of invitations sent in this method, and therefore the response rate, is unknown. The final dataset includes responses from both the initial distribution and this second distribution for a total of 302 librarians.

All of the respondents to the librarian survey work in comprehensive research-extensive institutions, while libraries in the libraries survey included associate, baccalaureate, and doctorate degree-granting institutions. Therefore, in order to remove a confounding factor from comparisons between the two survey results, only libraries at doctorate degree-granting institutions are included in this analysis. That way, although it is not known whether librarian respondents come from the same institutions that responded to the libraries survey, the official policies of academic research libraries can be better compared with the perceptions of librarians who work in that type of institution.

Questions to both library directors and librarians covered specific RDS offered or planned to be offered in their institutions, as well as opportunities for professional development on RDS issues for the professional staff. Half of these questions concerned informational or consulting RDS and half were about a greater level of involvement with technical/hands-on RDS. Informational/consulting services cover a wide range of services, from consulting on data management plans through discussing RDS with others:

- Consulting with faculty, staff, or students on data management plans.
- Consulting with faculty, staff, or students on data and metadata standards.
- Outreach and collaboration with other RDS providers either on or off campus.
- Providing reference support for finding and citing data or datasets.
- Creating Web guides and finding aids for data, datasets, or data repositories.
- Discussing RDS with other librarians, or other people on campus, or RDS professionals.

The technical or hands-on services show another level of involvement with RDS:

- Providing technical support for RDS systems (e.g., a repository, access, and discovery systems).
- Deaccessioning or deselection of data or datasets for removal from a repository.
- Preparing data or datasets for deposit into a repository.
- Creating or transforming metadata for data or datasets.
- Identifying data or datasets that could be candidates for repositories on or off campus.

Directly participating with researchers on a project (as a team member).

Library directors were asked whether each of the RDS were currently offered or planned to be offered in the future through the library. The answer choices were:

1. Not available, and we currently have no plans to offer it.
2. Not available, but we plan to offer it in more than 24 months.
3. Not available, but we plan to offer it within 13–24 months.
4. Not available, but we plan to offer it within 12 months.
5. Our library currently offers this service.

Librarians were asked how frequently they performed each of the RDS. Their answer choices were:

1. Never performed.
2. Performed a few times a year.
3. Performed about once a month.
4. Performed about once a week.
5. Performed daily.

In cases where the respondent indicated that service was provided to both faculty and students, separate details were given for faculty and students.

5. RESULTS

Providing reference support for finding and citing data or datasets is the most common of currently-offered or planned-to-be-offered informational RDS in academic research libraries, with nearly half currently offering this service and another third planning to within the next two years (Table 2). That means that almost 83% of these libraries will offer this service within the next two years. No other data informational service is currently offered by a majority of libraries, but consulting on data management plans, consulting on metadata creation, creating guides, and discussing RDS with patrons are planned in a majority.

Table 2. Informational or consulting RDS currently offered by the library or planned to be offered in the future [library study].

RDS	No plans	>24 months	13-24 months	<12 months	Has service
Discuss RDS (n = 99)	37.4%	11.1%	7.1%	16.2%	28.3%
Create guides (n = 99)	28.3%	13.1%	13.1%	20.2%	25.3%
Find datasets (n = 99)	17.2%	8.1%	14.1%	11.1%	49.5%
Outreach (n = 97)	53.6%	10.3%	8.2%	11.3%	16.5%
Consult meta (n = 98)	42.9%	10.2%	13.3%	10.2%	23.5%
Consult DMP (n = 99)	45.5%	11.1%	8.1%	9.1%	26.3%

Outreach and collaboration with other RDS providers either on or off campus is the least commonly offered information service now and least likely to be in the planning stages, although this may merely mean that the library doesn't need to collaborate in order to offer RDS.

The following two tables (Table 3A and Table 3B) show how often informational RDS are performed by librarians in academic research libraries for faculty (Table 3A) or for students (Table 3B). On average, informational RDS are currently performed by these librarians never or only a few times a year, with helping faculty find relevant data or datasets the most frequently offered service. Helping students find data or datasets is the most frequently offered RDS for students, followed closely by creating library guides to data services. Table 3A and Table 3B also show that in each case in which a service is performed for faculty or students, it is performed more often for faculty than for students.

Table 3A. Informational or consulting RDS currently performed by librarians for faculty [librarian study].

RDS	Never	Few times/ year	Once/ month	Once/ week	Daily
Create guides (n = 262)	44.0%	38.0%	10.0%	6.0%	2.0%
Find data (n = 255)	32.0%	41.0%	16.0%	6.0%	5.0%
Consult meta (n = 261)	52.0%	35.0%	7.0%	5.0%	1.0%
Consult DMP (n = 265)	55.0%	29.0%	12.0%	3.0%	0.0%

Table 3B. Informational/consulting RDS currently performed by librarians for students [librarian study].

RDS	Never	Few times/ year	Once/ month	Once/ week	Daily
Create guides (n = 239)	53.0%	33.0%	8.0%	4.0%	2.0%
Find data (n = 233)	36.0%	39.0%	11.0%	9.0%	5.0%
Consult meta (n = 240)	74.0%	20.0%	3.0%	2.0%	1.0%
Consult DMP (n = 239)	81.0%	13.0%	5.0%	0.0%	1.0%

Although over half of all research libraries in the libraries survey do not officially consult with others on campus or beyond the campus for RDS, individual librarians are slightly more likely to report that they collaborate on RDS (Table 4). Almost half of librarians engage in outreach, that is, collaborating with other RDS providers on campus. Working beyond their campus is more common, and 60% of the librarians report they participate in working groups or other professional groups about RDS. This participation may lead to a growth in RDS in libraries in the future.

Table 4. Informational or consulting RDS currently performed by librarians with others [librarian study].

RDS	Never	Few times/year	Once/month	Once/week	Daily
Outreach (n = 218)	53.0%	34.0%	8.0%	5.0%	1.0%
Groups (n = 218)	40.0%	35.0%	15.0%	8.0%	2.0%

In general, technical RDS are currently less frequently offered in libraries than are informational or consulting RDS, but only slightly so (Table 5). Roughly half of these academic research libraries offer or plan to offer most of the other technical services within two years, including (in order of frequency) within two years: providing technical support for a data repository, identifying datasets to incorporate into an institutional repository, providing librarians to serve as team members on e-science projects, creating metadata for datasets, and preparing data for deposit. Only deaccessioning datasets is not offered or planned by a majority of academic research libraries.

Table 5. Technical RDS currently offered by the library or planned to be offered in the future [library study].

RDS	No plans	>24 months	13-24 months	<12 months	Has service
Team member (n = 100)	49.0%	11.0%	5.0%	8.0%	27.0%
Identify data (n = 98)	44.9%	10.2%	12.2%	15.3%	17.3%
Create meta (n = 97)	50.5%	10.3%	16.5%	6.2%	16.5%
Prepare data (n = 100)	52.0%	9.0%	13.0%	11.0%	15.0%
Deaccession (n = 100)	73.0%	8.0%	10.0%	6.0%	3.0%
Tech support (n = 100)	43.0%	12.0%	16.0%	11.0%	18.0%

Librarians also report that they perform technical RDS less frequently than informational or consulting RDS, although the questions were slightly different in the librarian survey. Neither identifying datasets for inclusion in a repository nor serving as an e-science team member for faculty (Table 6A) or for students (Table 6B) is offered by a majority of these librarians. For those who do offer the services, a "few times per year" is the mostly likely frequency.

Table 6A. Technical RDS currently performed by the librarian with faculty or staff [librarian study].

RDS	Never	Few times/ year	Once/ month	Once/ week	Daily
Identify data (n = 256)	52.0%	31.0%	12.0%	4.0%	1.0%
Team member (n = 256)	65.0%	26.0%	5.0%	3.0%	1.0%

Table 6B. Technical RDS currently performed by the librarian with students [librarian study].

RDS	Never	Few times/ year	Once/ month	Once/ week	Daily
Identify data (n = 233)	71.0%	21.0%	4.0%	3.0%	1.0%
Team member (n = 231)	82.0%	13.0%	3.0%	1.0%	1.0%

Table 6C shows the technical services offered by librarians regardless of for whom. Very few librarians offer these services, although they may be covered by a single data services librarian in those institutions that support the service. Creating metadata for data is the most commonly offered service, but most frequently only a few times per year. There is wide variation: 207

of 226 respondents reported never performing deaccession or deselection of data or datasets from repository, although one respondent reported performing this service daily.

Table 6C. Technical RDS currently performed by the librarian on data or datasets [librarian study].

RDS	Never	Few times/ year	Once/ month	Once/ week	Daily
Tech support (n = 224)	76.0%	15.0%	4.0%	2.0%	4.0%
Deaccession (n = 226)	92.0%	7.0%	1.0%	0.0%	0.0%
Prepare data (n = 228)	71.0%	21.0%	6.0%	1.0%	2.0%
Create meta (n = 229)	67.0%	20.0%	6.0%	3.0%	4.0%

The library study also explored whether libraries provided opportunities for staff to develop skills related to RDS. Just under one-third (31 of 99) replied that they provide some opportunities. Those that replied yes also reported which of the following opportunities were provided:

1. In-house staff workshops or presentations.
2. Taking courses related to RDS.
3. Attending conferences or workshops elsewhere related to RDS.

Library directors could select all that applied to their organizations. Providing conference opportunities was the clear favorite, with 94% of libraries offering attendance at conferences to their librarians (Table 7). Slightly more than half of libraries reported supporting attendance at courses related to improving RDS skills.

Table 7. Percentage of libraries providing specific opportunities for staff to develop RDS skills. Library directors were allowed to select all that applied [library study].

Opportunities for RDS skills (n = 99)	Percentage
Training	45%
Courses	58%
Conferences	94%

Among libraries that currently offer any training opportunities, Table 7 shows the percentages of libraries offering each opportunity. However, when all libraries are considered together, including those that do not offer services, the percentages are much lower: 29% for conferences, 18% for courses, and 14% for in-house training.

Some librarians feel they have opportunities for learning. Just under half (47%) of the 219 librarians who answered this question felt they had the opportunity for at least one type of RDS skills development (Table 8). The most common opportunity was support for attendance at conferences (65%), followed by courses elsewhere (53%), and, less often, training at their library (32%).

Table 8. Percentage of librarians agreeing that their library provides specific opportunities to develop RDS skills [librarian study].

Agreement with opportunities for RDS skills	Percentage
Training (n = 218)	32%
Courses (n = 216)	53%
Conferences (n = 217)	65%

6. DISCUSSION

The most commonly offered or planned informational RDS, finding and citing datasets, (Table 2) is a service that simply extends a familiar library reference service into the realm of data. At the other extreme, the least commonly offered information service is outreach and collaboration with other RDS providers. With the extensive hardware, software, and educational components needed for effective RDS, it is somewhat disheartening that so few research libraries are collaborating with others.

According to librarians who work in academic research libraries, RDS are being performed never or only a few times a year (Table 3A and Table 3B). Keeping in mind thatTable 2 takes into account planned availability as well as current availability, this result is not surprising. Growth in current performance of RDS by librarians can be expected to follow growth in current availability of RDS by libraries.

Considering technical RDS, this type of service is less available in libraries than are informational RDS (Table 4). The picture is also likely to change significantly in the near future, as only one service (deaccessioning or deselection of data or datasets for removal from a repository) is neither in the plans nor currently being offered by most libraries. Perhaps the preservation aspect of repositories is considered inconsistent with deselecting data that is put into a repository.

There appears to be somewhat of a mismatch between what academic research library directors believe they offer to their librarians and what the librarians themselves perceive to be available to them in the way of RDS training opportunities (Table 8). Nonetheless, these results portend well for the future of

RDS, as there are clearly some opportunities for training of librarians in RDS skills.

Library directors and librarians who are aware of research data management issues or are currently involved in RDS are more likely to respond to the survey, so results may show an inflated picture of research library involvement in RDS. Also, this is a dynamic topic and plans may change at some of the responding institutions. Responses to the libraries survey concern library policy, with the unit of analysis at the library institutional level. Responses to the librarian survey concern individual perceptions and opinions, with the unit of analysis at the individual librarian level. It is not possible to know how many of the librarian-respondents work at the institutions that responded to the libraries survey. It is also not possible to know how many individuals from the same institution responded. In addition, since this paper focused on libraries and librarians at research or doctorate-granting institutions, the results are not generalizable to academic libraries and librarians at other types of schools such as baccalaureate and associate degree-granting institutions. And, finally, as with any survey, responses are self-reported and are assumed to be accurate and truthful, although this cannot be verified.

7. CONCLUSION

It is clear that some academic research libraries are offering a variety of research data management services and more plan to do so within the next two years. Most commonly these services are extensions of traditional informational or consultative services, such as helping faculty and students locate datasets or repositories. A small, but growing, number of libraries are becoming more involved with research data, from helping with data management plans to preparing and preserving research data for deposit in data repositories.

Many of the librarians who work in academic research libraries feel they have the subject knowledge necessary to help their constituents with research data services, but need the opportunity to take advantage of continuing education. Whether consultative or hands-on services, librarians need opportunities to learn more about these services either on their own campus or through attendance at workshops and professional conferences.

Working with others on campus, as both teachers and joint learners of research data service specifics, will help the library play a shared role in building the future of research data at their universities.

The comparisons drawn here between library policy on RDS and the perceptions of front-line librarians as they implement this policy, indicates some misalignment. However, that is to be expected, as most libraries are in the early stages of making RDS available. Increased awareness of this issue within the academic library community is likely to result in more effective development of RDS.

ACKNOWLEDGMENTS

This work was supported by Data Observation Network for Earth (DataONE), National Science Foundation award #0830955 under a Cooperative Agreement. We would like to thank Graduate Research Assistant Madison Langseth for her careful work on helping us respond to the peer reviewers' suggested revisions.

REFERENCES

1. Association of Research Libraries (2006). To stand the test of time: Long-term stewardship of digital data sets in science and engineering. Washington, D.C.: Association of Research Libraries. Retrieved from http://arl.nonprofitsoapbox. com/storage/documents/ publications/digital-data-report-2006.pdf

2. Association of Research Libraries (2010). E-science and data support services: A study of ARL member institutions.Washington, DC: Association of Research Libraries. Retrieved from http://www.arl.org/storage /documents/publications/ escience-report-2010.pdf

3. Bach, K., Schafer, D., Enke, N., Seeger, B., Gemeinholzer, B., & Bendix, J. (2012). A comparative evaluation of technical solutions for long-term data repositories in integrative biodiversity research. Ecological Informatics, 11, 16–24.

4. Brown, E. (2010). I know what you researched last summer: How academic librarians are supporting researchers in the management of data curation. The New Zealand Library & Information Management Journal, 52(1), 55–69.

5. Brown, S., & Swan, A. (2007). Researchers' use of academic libraries and their services: A report commissioned by the Research Information Network and the Consortium of Research Libraries. London, UK: Research Information Network and Consortium of Research Libraries in the British Isles.

6. Carlson's, J. (2012). Demystifying the data interview: Developing a foundation for reference librarians to talk

with researchers about their data. Reference Services Review, 40(1), 7–23.

7. Cheek, F. M., & Bradigan, P. S. (2010). Academic health sciences library research support. Journal of the Medical Library Association, 98(2), 167–171.

8. Corrall, S., Keenan, M. A., & Afzal, W. (2013). Bibliometrics and research data management services: Emerging trends in library support for research. Library Trends, 61(3), 636–674.

9. Council on Library and Information Resources (2008).No brief candle: Reconceiving research libraries for the 21st century. Washington, DC: Council on Library and Information Resources.

10. Creamer, A., Morales, M. E., Crespo, J., Kafel, D., & Martin, E. R. (2012). An assessment of needed competencies to promote the data curation and management librarianship of health sciences and science and technology librarians in New England. Journal of eScience Librarianship, 1(1), 18–26.

11. Gabridge, T. (2009). The last mile: The liaison role in curating science and engineering research data. Research Library Issues: A Bimonthly Report from ARL, CNI, and SPARC, 265. (pp. 15–21).

12. Gold, A. (2007). Cyberinfrastructure, data, and libraries, part 2: Libraries and the data challenge: Roles and actions for libraries. D-Lib Magazine, 13(9/10). Government of Canada (2013). Tri-agency open access policy.

13. Hey, T., & Hey, J. (2006). E-science and its implications for the library community. Library Hi Tech, 24(4), 515–528.

14. Hey, T., Tansley, S., & Tolle, K. (2009a). The fourth paradigm: Data-intensive scientific discovery. Redmond, WA: Microsoft Corporation.

15. Hey, T., Tansley, S., & Tolle, K. (2009b). Jim Gray on eScience: A transformed scientific method. In T. Hey, S. Tansley, & K. Tolle (Eds.), The fourth paradigm: Data-intensive scientific discovery (pp. xix–xxxiii). Redmond, WA: Microsoft Corporation.

16. Jones, E. (2009). Reinventing science librarianship: Themes from the ARL-CNI forum. Research library issues: A bimonthly report from ARL, CNI, and SPARC, 262. (pp. 12–17).

17. Kuipers, T., & Van der Hoeven, J. (2009). Insight into digital preservation of research output in Europe: Survey report (D3.4). Didcot, UK: PARSE.Insight. MacColl, J. (2010). Library roles in university research assessment. LIBER Quarterly, 20(2), 152–168.

18. Markauskaite, L., Kennan, M. A., Richardson, J., Aditomo, A., & Hellmers, L. (2012). Investigating eResearch: Collaboration practices and future challenges. In A. Juan, T. Daradoumis, M. Roca, S. Grasman, & J. Fauli (Eds.), Collaborative and distributed e-research: Innovations in technologies, strategies and applications (pp. 1–33).

19. Newton, M. P., Miller, C. C., & Bracke, M. S. (2010). Librarian roles in institutional repository data set collecting: Outcomes of a research library task force. Collection Management, 36(1), 53–67.

20. Peters, C., & Dryden, A. R. (2011). Assessing the academic library's role in campus-wide research data management:

A first step at the University of Houston. Science & Technology Libraries, 30(4), 387–403.

21. Potter, W. G., Cook, C., & Kyrillidou, M. (2011). ARL profiles: Research libraries 2010. Washington, D.C.: Association of Research Libraries.

22. Steinhart, G., Saylor, J., Albert, P., Alpi, K., Baxter, P., Brown, E., et al. (2008). Digital research data curation: Overview of issues, current activities, and opportunities for the Cornell University Library. A report of the Cornell University Library Data Working Group.

23. Tenopir, C., Birch, B., & Allard, S. (2012). Academic libraries and research data services: Current practices and plans for the future. (An ACRL white paper). Chicago, IL: Association of College and Research Libraries.

24. Tenopir, C., Sandusky, R. J., Allard, S., & Birch, B. (2013). Academic librarians and research data services: Preparation and attitudes. IFLA Journal, 39(1), 70–78.

25. University of Minnesota Libraries (2011). Funding agency and data management guidelines. Minneapolis–St. Paul, MN: University of Minnesota.

2

CHAPTER

Where are We going? Overview of Major Collaborative Conservation Research Projects and Implications for the Future Preservation of Collections

Jonas Palm

National Archives, Department of Preservation, PO Box 12541, 102 29 Stockholm, Sweden

ABSTRACT

In this paper I will tell you about some research and strategic projects, some of them at the end of their time and some just started. Although digitisation seems to have been a major topic in archives and libraries since the mid 1990s the vast majority of the collections are still based on traditional, i.e. analogue, media. The preservation problems much discussed prior to this still exist and still present some major hazards. What differs today from yesterday is that the technological progress has given us new means within reasonable costs to perform research and develop new tools as well as improve the monitoring of decay.

INTRODUCTION

In this paper I will tell you about some research and strategic projects, some of them at the end of their time and some just started. Although digitisation seems to have been a major topic in archives and libraries since the mid 1990s the vast majority of the collections are still based on traditional, i.e. analogue, media. The preservation problems much discussed prior to this still exist and still present some major hazards. What differs today from yesterday is that the technological progress has given us new means within reasonable costs to perform research and develop new tools as well as improve the monitoring of decay.

A general tendency is also to find solutions, which may be applied on larger groups of items - and preventive measures rather than emergency measures - since the costs for conservation may be quite substantial for one single object. It is necessary to find cost effective strategies, which comply with the task to preserve the cultural heritage.

What then are the preservation problems that are still persistent? When it comes to paper based media it is still the deterioration of paper and iron gall ink corrosion. Brittle paper is still a major cause for concern even though modern paper quality has improved to some extent. Parchment and its true properties are still to some extent unknown for most of us. The medium has been looked upon as almost indestructible but recent findings show we may have to take another look at this.

A major problem with studies of deterioration, have been the difficulties to find appropriate accelerated ageing tests to simulate natural ageing and to check whether the results correspond to the slow real time deterioration processes. There hasn't been any economically feasible means to measure deterioration at natural ageing speed. Analogue audio- and videotapes present a major economic problem in the near future, if not already.

How to preserve digital image files - commonly known as digital images - is a major issue these days as the problems and costs become more and more evident. This field of media has a lot of economic potential as well as it is an open field for new inventions. Thus the ways and strategies of preservation to choose from may change quickly. The need for an interface of some sort - both hardware and software - between the information and the beholder, makes these issues extremely difficult to tackle. All these issues will continue to initiate research as well as results. They will, of course, also initiate new questions.

THE INKCOR PROJECT

Iron gall ink was probably the most widely used ink in the Middle Ages. It was used on both parchment and paper. A huge number of documents still exist in pristine condition but maybe an equally huge number suffer from - or will suffer from - iron gall ink corrosion. The problem of iron gall ink corrosion was, as far as I remember, first looked upon scientifically in Germany in the 1970s. Also copper pigments were studied at the time but only as an ingredient in blue or green colours and not in ink. De-acidification of some kind was considered as the solution.

At the end of the 1990s the thread was taken up again at a meeting in Rotterdam, which eventually resulted in the Fifth Framework EU-project InkCor - Stabilisation of iron-gall ink containing paper. The approach now was to use anti-oxidants as the remedy.

In order to make it possible to do research on the effects of stabilisation agents on material more closely related to reality than the use of modern papers and inks, the InkCor project collected numerous historical ink recipes - now available in the ICN (Instituut Collectie

Nederland) Art Technological database - and produced historical model papers.

The properties of old inks on model paper were studied in order to obtain a better knowledge of how and why differences in deterioration occurred seemingly irrelevant of age. In-air Proton Induced X-ray Emission analysis - in short PIXE analysis - combined with the measuring of some basic properties of paper resulted in the conclusion that the stability of historical iron gall ink containing papers can be estimated with information non-destructively obtained.

Research on the problems of iron gall ink has for decades been focused on the iron content of inks. InkCor-produced results show that this was probably a simplification of the problem. Copper has long been known to be a very corrosive substance and the project has now shown it to be another major component in historical inks. As it was shown that phytate treatment - an anti-oxidant - mainly reacts with iron ions the need to find an effective anti-oxidant for the stabilisation of other metal ions is obvious. A simple identification test for iron has been developed by ICN in the Netherlands and within the InkCor a new one, aimed at identification of copper ions was developed.

Since there have been indications that gelatine may have a delaying effect on the iron gall ink corrosion of paper this well known substance is more and more implemented as a non controversial procedure by conservators. To establish whether this was an effect on the paper or the iron gall ink, an experiment was performed to evaluate the impact of gelatine as a sizing agent on the degree of polymerisation of cellulose. Samples were prepared with a lab-produced ink and some of these were afterwards immersed in water and others were treated with gelatine in three different ways - immersion, spraying and application with brush. After ageing gelatine was washed out of the samples, size exclusion chromatography (SEC) was used to determine polymerisation. No significant effect of ageing stability on iron gall ink containing paper was found.

Results from the InkCor-project have resulted in several improvements of an existing commercial mass de-acidification process implemented by Zentrum für Bucherhaltung in Germany. The impact of the results from the InkCor- project on how to deal with iron gall ink corrosion is substantial. They will also have some impact on the new PaperTreat-project, which will start this year.

THE PAPERTREAT PROJECT

PaperTreat - evaluation of mass de-acidification processes - is an extensive development of the research, which was done on mass de-acidification in the 1980s and 90s. In the early 1990s digitisation blacked out most preservation issues. Now things seem to have settled down to where they probably belong and the EU's 6th framework programme has decided to fund this project.

The strategic objects are

- to evaluate the impact of modern material and techniques on historic heritage materials,
- to undertake a comparative assessment of traditional and modern protection and conservation strategies.

The issue of brittle paper in books and documents has long been established and today about 25% of the books in general library collections are brittle and an additional 60% endangered. A variety of mass deacidification techniques have been used for more than two decades and have been shown to prolong the lifetime of treated paper by up to 300%. Mass deacidification has also been proven as a cost-effective strategy, with microfilming being about 8 and digitalisation 30 times more expensive. Consequently, mass deacidification has been incorporated into the preservation strategy of numerous institutions in the USA. Despite the recognized need for mass deacidification across Europe, it is used only by a few institutions, while numerous others plan to perform an evaluation of existing treatments, prior to opting for one.

A number of comparative evaluations of the traditional treatments had been performed in the past, however, 5 new ones were developed recently.

The specific aim of the project is to

- identify side effects of the treatments;
- enable estimation of the extension of the useful life of paper, as achieved by traditional and new treatments;
- provide cost specification for each treatment.

The results will enable a sound decision on the type of treatment best suited for certain materials.

In addition to comparative analysis of the traditional and novel mass deacidification techniques, the project will develop:

- standard model materials;
- standard evaluation criteria;
- standard quality control criteria, which will allow for superior comparative assessment of new processes with the traditional ones and considerably simplify the evaluation of emerging new technologies in the future.

THE IDAP PROJECT

Parchment in the form of manuscripts, scrolls, charters, book covers, and substrates for artworks, belong to the most valuable objects of European cultural heritage for they are the bearers of the history, which has shaped the development of European society over a period of many centuries. Large collections exist in public and private libraries, archives, and museums and in diverse religious foundations in varying degrees of preservation. The IDAP-project, yet another EU-funded preservation project, is developing improved damage assessment including an early warning system based on a proven analytical protocol which would contribute to extending the lifetime of these documents as well as improving methods used for their conservation. The aim is to implement procedures for improved damage assessment and preservation of historical parchment. Currently, the condition of parchment documents is assessed by conservators-restorers in terms of their visual appearance and mechanical strength. Deterioration (biological, chemical, physical), however, takes place at several levels and damage cannot always be readily visually assessed. This then poses a complex problem of preservation. That this is indeed true is shown by the results of statistically randomly sampled documents at the National Archives in Stockholm. A collection of 20,000 parchment documents had been considered as being in very good condition. The assessment showed different as quite a few were in very bad condition. The concluding report will make very interesting reading for those involved with the preservation of parchment documents. The project will finish in August 2005.

The Papylum Project: Chemiluminescence

For ages the task of determining the condition of a paper-based object has been based on a mixture of visual inspection and tests for which most of them require a physical sample taken from the original. The use of spectroscopic analyses and the introduction of a new tool to study chemiluminescence have taken studies of deterioration of paper to a new level. Not only is it now possible to study the physical properties of paper but also the chemical activities of oxidation.

Within the Papylum-project an instrument has been developed which makes it possible to study chemiluminescence during the oxidation of paper. Chemiluminescence is the weak light emitted by molecules as a result of chemical reaction. This means that studies of paper degradation can be studied at room temperature and seen virtually. The use of this method has among other things resulted in

- new knowledge on cellulose oxidation and the influence of carbonyl groups,
- development of a software which can be used to predict rates of degradation at room temperature,
- guidelines for conservation researchers on how to conduct studies of the photo-stability of paper,
- relative humidity apparently having an important role in the deterioration of moderately alkaline papers.

A study, emphasising the promising results above, on the correlation between polymerisation as measured with chemiluminometry and double fold - a traditional well established paper test - was presented. Observed correlations showed that further investigations into the use of chemiluminometry as an analysis tool to be used on historical papers ought to be carried out.

The SurveNIR Project

As a continuation of Papylum this year the EU-funded project SurveNIR will start. Surveying of collections is usually done using simple physical and chemical tests, often destructive, in order to reveal the collection condition, the general conservation needs and to plan preservation activities. As the testing of a single item should take as little time as possible, the information obtained is usually only rudimentary.

By providing a near-infrared-spectroscopic instrument prototype, the end-users will be provided with a powerful survey tool allowing reagent-less multi-component low-cost analysis of items. The technique itself has recently found its way into industry as an in-line production control tool.

The spectra can be taken non-destructively in a matter of minutes, then analysed using advanced statistical methods and, as a result, data important for the preservation officer and policy maker may be provided, such as the material composition and chemical state, via a user-friendly software interface.

The result of the project will be improved tools to e.g

- determine the condition of a collection,
- assess future needs,
- identify major causes of deterioration,
- identify vulnerable areas of the collection,
- facilitate management and resources.

The tools will be tested in seven collections, thus demonstrating the relevance of the developed approach. The involvement of a SME, a university lab and two libraries, three archives, and two museums, i.e. seven end-users, ensures an efficient and directed dissemination both of knowledge and of the developed technology.

Hyperspectral imaging

As we are in the realm of light I'd like to inform you of an interesting project underway in the National Archives in the Netherlands. It is a new way to use scanning within preservation. Non-destructive optical techniques have always belonged to the most important investigation methods applied in paper and writing durability research. Visual inspection of a document is a fast and inexpensive way of detecting critical areas, where for example ink corrosion occurs in an advanced state, or where discolouration or staining jeopardizes the legibility. From visual inspection alone, the experienced researcher can get a qualitative impression of the general condition of the inspected material.

However, for comparing the state of conservation of documents with each other without actually having to view them simultaneously, quantitative analytical methods are required. Besides a number of other sophisticated spectroscopic techniques, optical reflectance spectroscopy and colour measurement, has been used successfully for analysing local material properties in relation to, for example, paper degradation. Unfortunately, practically all of the quantitative optical techniques are per se non-imaging, i.e. each measurement represents the average properties only around a single point on the document.

The Dutch project has taken another approach with hyperspectral imaging (HIS). An advanced multi-spectral digital imaging system gathers optical information from a significant document area with a high spatial resolution, and can thus overcome the problems typical for single-point techniques. The term hyperspectral imaging refers to the acquisition of a series of digital images at a large number (50-100) of different, well-defined optical wavelengths in the ultra-violet, visible and near infrared. HSI results in the simultaneous, precise determination of the reflectance spectra from all areas of a document with a high spatial resolution. The reflectance spectra often show significant differences for different types of ink (or other substances), which may look identical in direct viewing or conventional RGB imaging. Using dedicated software, the spectral properties can be exploited for distinguishing inks or areas exhibiting different degrees of paper degradation, as well as for the automatic mapping of their distribution.

This kind of equipment can be used to take "finger prints" of objects in order to study changes over time. Except for the usual studies of deterioration or other changes it could be interesting to study objects after being on display. Items on loan can be fingerprinted before being sent off and checked when they return. The possibilities are numerous. And the best of all: yet another non-destructive analysis tool.

These are just a few of the examples of research in the area of preservation. A lot of other research projects on a more local level are in progress. These include further studies in accelerated ageing and the development of non-destructive analysis tools.

AUDIO AND VIDEO

Analogue audio- and videotapes are potentially as great a problem as brittle paper. Millions of hours are recorded on analogue magnetic tapes, which slowly deteriorate. Magnetic information changes over time creating losses in audio and video quality. Acetate has been used as a carrier and the binding layer between carrier and magnetic layer may deteriorate with disastrous results of course. The only way to preserve this kind of information for the future is to digitize it. Actually they are the only media, which have to be digitized to have a chance to be preserved for perpetuity. This means a long-term commitment, which will cost a lot - I repeat: a lot - of money. Both for the initial transferring process, and for the continual merging from generation to generation of computer systems. Thus it is only realistic to assume that any measures taken will only be made by major institutions, probably governmental bodies like national archives and libraries.

THE TAPE PROJECT

To cope with the numerous small collections held outside the major national institutions the TAPE-project was started in 2004. TAPE (Training for Audiovisual Preservation in Europe), funded under the Culture 2000 programme of the EU, is a 3-year project for raising awareness and training by expert meetings, research, publications and workshops. TAPE focuses on audiovisual collections held outside the major national institutions with specific responsibilities for audiovisual heritage. Through a programme of activities TAPE aims to explore the requirements for continued access to audiovisual materials and the application of new technologies for opening up collections that provide living documentation of the world of the 20th century.

Future life

The issues above emphasize preservation as I said in the beginning. Conservation is not forgotten though. At the British Library a meeting was held in September 2004 where delegates from both sides of the

Atlantic met to discuss the future of applied conservation research. A lot of ideas were put together in three priority groups:

1. Life cycle prediction, natural ageing of materials, evaluation of preservation strategies for different materials.
2. Effects of the storage environment and selection of the optimum environment for different materials.
3. Non-destructive methods for assessing damage to materials.

The British Library will take the lead in developing a coherent work programme by:

- circulating outline projects and establishing consensus with colleagues in the legal deposit libraries and national archives in the UK and Ireland, by the end of January 2005;
- fleshing these ideas out into a number of costed work strands with a clear timetable, initially for the period 2005 – 2010, by the end of April 2005;
- Involving collaborators in North America and Europe whose expertise and facilities complement those available in the UK and Ireland.

The project is restricted to paper, parchment and leather.

IS&T Archiving conferences

An area, which combines the traditional and modern media, is photography. We use digital cameras extensively, produce digital image files equally extensively and we talk about digital images. What is a digital image? It cannot be seen unless it is analogue. Thus we have a plethora of print outs - laser prints, inkjet prints, heat transfer prints, traditional photographic prints etc.- all made from digital image files. I will not go further into this but I recommend strongly that you follow what is happening within this field. Last year the Society for Imaging Science and Technology, IS&T, held their first Archival Conference in San Antonio, Texas. A second conference was held in Washington, DC, this year and the third will be held in Ottawa, Canada, next year. Here people meet from HP, Kodak, Fuji, national archives, museums, national libraries etc. and the papers presented are the latest things related to archiving digital image files and their offspring. The two first conferences were very informative indeed. The information is the latest on what has happened or is happening, even what is supposed to happen, and since this is a very dynamic field it is crucial for the archive and library community to follow suit with discussions and development, to be able to understand the processes and the paths they are taking.

This has been a fairly short summary of just some of the activities going on within the preservation community. Next year we will follow up on the events happening and we will present it in Uppsala.

WEB SITES REFERRED TO IN THE TEXT

ICN - Instituut Collectie Nederland. http://www.icn.nl/

IS&T - Society for Imaging Science and Technology. http://www.imaging.org/

Nationaal Archief. http://www.en.nationaalarchief.nl/

3

CHAPTER

Preparing for New and Changing Roles in Research Libraries – the Need for Continuing Professional Development

Gitte Larsen

Royal School of Library and Information Science, Dept. of Continuing Education and Consultancy, Birketinget 6, DK-2300 København S, Denmark

ABSTRACT

It is expected that library staff are qualified to offer high quality services to users visiting the physical library. Likewise, it is expected that they have substantial knowledge and skills needed for developing and maintaining electronic services and for dissemination of relevant services and facilities requested by the web-user. Serving remote library users calls for additional competencies, such as marketing, branding and communications skills in the electronic environment as well as knowledge of measuring and evaluation of the use of electronic services. It is a challenge to the staff to match particular needs and demands from different user groups, but also to library management staff to ensure that the competencies and skills are available in the organisation to match the needs of the user – wherever s/he might be located. Competencies are, in this context, defined as the combination of knowledge and experience that make the individual able to take the right actions in the daily working environment.

What education and training needs emerge from the changing roles and new tasks? How might we identify the needs for continuing professional development? And how can we maintain and update skills and competencies acquired maybe 25 years ago? These are key questions – not only to be addressed to library managers, but also to be considered carefully by those institutions responsible for continuing education and professional development of library staff.

THE NORDIC CONTEXT

The hybrid library, as it is developing today, combines the physical and the net-based library. The physical research library means the classical library: the collections, the study place for students and researchers, where the staff takes care of collection development and management, and serves the users with all kinds of materials and reference and enquiry services. An increasing amount of effort and resources in research libraries are, however, being spent on the provision of access to electronic resources and services. Today's challenges are content management, creation of institutional repositories, copyright negotiations, licence agreements and digital rights management. Qualified guesses from directors of Danish research libraries are that 50-65% of total staff resources go into net-based activities. The trend is similar in many of the larger public libraries. Previous working tasks, such as lending routines, are carried out as self-service, acquisition of new material is outsourced, etc.

Library staff is heavily involved in projects on developing e-services, creation of subject portals and gateways, and to customise access to e-resources to the individual faculty, to the individual researcher and student or to other target groups, but what influence do electronic library materials have on library loans and usage? Library statistics from Danish research libraries from 2004 show that 66% of library loans were 'down-loads' (print-outs of e-journal articles) and it is estimated that the use of electronic resources in Denmark's Electronic Research Library (DEFF) will increase by 25% per year during the next couple of years.

Research libraries are closely connected to their parent educational institutions. In the Nordic countries, universities are facing many challenges: the changing population of students (from elite to mass education), knowledge production in new forms (i.e. Open Archives); the demands for life long learning, and globalisation. These changes impact the ability of libraries to be an active player in knowledge production. Libraries have to create interfaces to the global knowledge system, tailored to the individual educational institution - the task is to import and create knowledge, to make knowledge available and provide information sources for the individual.

How might the development of the hybrid library and the change of tasks and roles of libraries influence performance planning for the entire staff? This is a key question - not only to be addressed to library managers, but also to be considered by those institutions responsible for continuing education and professional development of library staff. What kind of training programmes will we have to develop to support the challenges of the staff?

Core skills of librarians are still relevant - if not more relevant - in electronic information provision. These core skills include information handling, training and facilitating, evaluation and customer service. Core skills as cataloguing, classification (knowledge organisation), information retrieval, reference work and user education are fully applicable in an electronic context and all have a place in facilitating the effective use of internet in a changing information environment.

Emerging roles and competencies of librarians

Corrall has conducted a very rich literature study on the developing roles of librarians (Corrall, 2005). She refers to an American survey conducted in 2001, in which 947 academic, school and public librarians identified the most significant roles of librarians for the next five years as follows:

- instructing users in the navigation and evaluation of print and digital information;
- directing users to appropriate information resources;
- evaluating and purchasing resources for collection development;
- organising and cataloguing resources;
- creating programmes and services for under-served communities;
- creating new guides and/or navigational tools for electronic resources;
- archiving and digital preservation.

Except for the term 'digital' - like many other studies - it focuses on traditional roles of library staff only. Today, it is expected that library staff are capable of offering high quality services to the users who are visiting the physical library. Likewise, it is expected that they have sufficient knowledge and skills needed for developing and maintaining electronic services and for making available relevant services and facilities demanded by the web-user. Collection development and maintenance is as important as before, but in this context, new and revised strategies for e-resources acquisition have to be set up in the individual library, matching the user's needs and expectations. Libraries have to join consortia and other "buying clubs" and skills in contract negations are emerging.

Emerging skills in the electronic information environment are communication, negotiation and collaboration. Collaboration on 'Ask a Librarian' virtual reference services, such as the Danish Biblioteksvagten (a consortium with participation from 44 public libraries and 13 academic libraries working in turns) have demonstrated that the ability to co-operate virtually across institutions and the ability to accumulate experiences and knowledge sharing are important personal skills.

Marketing, branding and communication skills are important additional competencies amongst library staff in the physical library. Serving remote library users calls, however, for additional competencies in using suitable techniques for meeting the user in the electronic environment, such as direct marketing, chat, communication via SMS and weblogs, writing to the Web and Web design.

In order to meet the needs and expectations of the remote library user, knowledge of measuring and evaluation of the use of electronic services (usability testing) is important. Do we know who the users really are? Do we know what they expect and whether they are satisfied with the services provided by the library or not?

Buying access to electronic library materials is very expensive, but mandatory. Journals, directories, textbooks, newspapers, music, sheet music and other materials are being digitised and made available in e-formats. A key issue in collection management for libraries is therefore to monitor the budget for e-resources to have it spent cost-effectively. Use statistics and performance management for electronic journals are some of the highly relevant issues that library staff is facing right now.

It is a challenge for the staff to match the particular needs and demands of different user groups and for library management staff to ensure that the competencies and skills in the organisation are able to match the needs of the user - wherever s/he might be located.

Development of knowledge, skills and attitudes

Competencies are, in this context, defined as the combined theoretical knowledge and practical experience that makes the individual able and willing to take the right decisions in the daily working environment. This definition is important, because it emphasises that competencies are not equal to professional formal qualifications only. The figure below shows a general model of the interrelation between human resources and behaviour - and explains why continuing professional development and training is important for the further development of organisations.

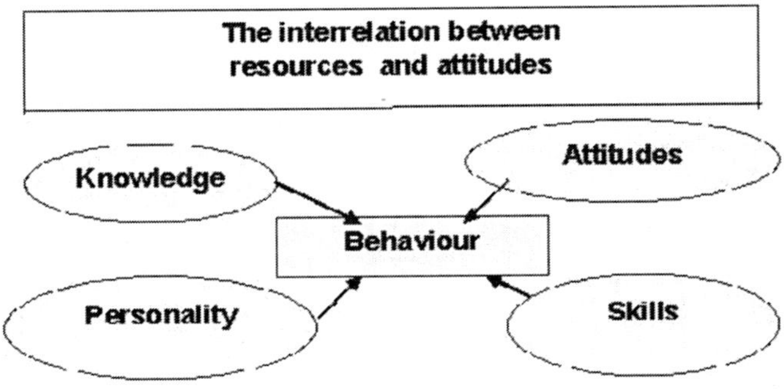

Continuing professional development in a Danish context

The Royal School of Library and Information Science (RSLIS) states in its performance contract with the Ministry of Culture 2003-2006, that we aim to consolidate the school in its position as the most important institution for continuing education and training within LIS and related areas relevant for library development. Further the contract says that (a part of) our training courses should underpin the aim of strategic library development as expressed in the Danish Library Act from 2000, which has a paragraph on national strategies for development of new competencies in public libraries.

On an annual basis, we offer for all kinds of library staff a selection of approximately 200 different training courses and seminars in addition to formal continuing education, such as our Master Programme in Library and Information Science. Each semester, some 30-40 new course titles are introduced - often developed on suggestions from libraries. Topics from training courses published in our course catalogues are used as a basis for tailoring courses, seminars, workshops or talks for individual libraries, organisations or interest groups.

The Department of Continuing Education and Consultancy invites all the academic staff from RSLIS (approximately 70 full time staff) to participate in our activities. Further specialists from practice with special subject knowledge and freelance consultants are involved in the training.

In order to stimulate new skills and competencies, RSLIS offer a selection of opportunities:

New knowledge: a Master of Library Science Programme as a postgraduate programme for those with a degree in LIS or an equivalent background. It leads to the award of the qualification of MLISc. This programme is taught in English. It comprises 4 modules and one special subject module. Each module is of one semester's duration, corresponding to 30 ECTS. The 4 modules are:

- Information seeking and information retrieval theories.
- Knowledge organisation.

- Knowledge management.
- Dissertation, which can include cultural mediation, children's culture and library management.

RSLIS offers in addition a part-time Masters Programme for librarians with at least 2 years of practical working experience. The Masters programme has been very successful for librarians who want to have formal continuing education in LIS. RSLIS is, of course, not the only Danish provider of continuing education of relevance for library staff. Programmes offered by Danish universities such as Masters degrees in Public Management, Public Administration and in Public Policy, have attracted quite a few librarians.

During the last couple of years, the majority of Danish public library directors have updated their formal qualifications in professional management. Since public administration becomes ever more complicated, a large scale programme was set up to offer a Diploma Degree in library management with financial support from the state.

In RSLIS, training courses are developed in order to cover the following needs:

- New practical skills (qualifications and personal skills).
- Updating practical skills (qualifications and personal skills).
- New inspiration.
- 'Hot topics' - orientation on ongoing policy and legal issues.
- Development of the organization
    - the individual staff member,
    - the team / the unit,

- the management level,
- the entire organisation.

The current development needs of the hybrid library as mentioned above are supported by the first three categories of training courses.

Governmental policy issues often generate 'hot topics' by changes in laws and regulations. For instance, when the Danish Minister of Culture proposed that all public libraries should install net filters to protect children from pornography or when the government decided on a radical change of the local authority structure (a decline from some 270 municipalities to 98 from 2007). This kind of change in local governance means that libraries have to merge and set up new structures for IT, management and work processes. In RSLIS we aim to pick up and influence that kind of library issue - typically in seminars and conferences. Library organisations such as the Danish Library Association (DBF) and The Danish Research Library Association (DFDF) will of course offer conferences on 'hot topics' of relevance for their members - but with more emphasis on the political and practical perspective.

During the last couple of years an increasing number of training courses have dealt with organisational behaviour. RSLIS organises practical training, seminars and workshops focusing on organisational development, which has been one of the most demanded topics. Typical themes are: scenario planning, establishment of common values, cooperation issues, conflict handling, motivation, coaching, stress management, team building and senior staff. Popular themes on the further development of the individual are: self-management, efficient work planning and how to get new ideas.

New recruitment: competencies: knowledge, skills and attitudes

On new recruitment, library directors want staff with a deep (and broad) theoretical educational background, combined with strong personal and social competencies. What do these demands indicate and how will they influence existing staff and recruiting policy?

A research project carried out in 2002 in the Queensland University of Technology in Australia aimed to identify and examine skills and knowledge essential for the successful library and information professional in the twenty-first century. This study considers the dual dimension of discipline knowledge and personal attributes, or generic capabilities.

Discipline knowledge includes, in addition to relevant subject knowledge, also career planning skills and ability to do research in LIS.

Generic capabilities include personal attitudes, such as being capable of lifelong learning, teamwork, communication, project management, critical thinking, problem solving, business acumen and self-management.

In the UK in 2002 the Chartered Institute for Library and Information Professionals (CILIP) presented a set of professional information skills and generic capabilities based on analysis of job descriptions, interviews and research reports.

Professional information skills:

- Knowledge management: information architecture, ICT skills, technical (traditional) skills.
- Subject expertise: collection management; collection description, technical (traditional) skills.
- Information technology: design, application, systems, user support (problem solving).

- Service development: user information, survey, service impact analysis, planning and evaluation, promotion and marketing.

Generic skills:

- Project management: people management, research skills, bids and proposals.
- Critical skills: thinking, analysis, problem solving, research.
- Leadership: generic management, communication skills, strategic management, people skills, financial skills.
- Promotion and marketing: presentation skills.

A study amongst 53 US library directors of academic libraries provides us with an interesting profile for recruitment of new staff for the libraries at the start of the 21st century (Bridges, 2003). When expressing their wishes on the educational background of new staff it is interesting to note that they are demanding a solid broad education - not a specific one. Personal skills and attitudes as expressed in the figure below were considered more important than professional qualifications.

Knowledge	Skills	Attitudes
• a solid broad education • knowledge of technology • 'the ability to tell the difference between new technology and necessary technology'	• customer service skills • teaching skills • creativity • ability to market library services • interpersonal ability • drive to learn • communication skills • ability to think conceptually	• intellectual curiosity • risk taking • sense of humour • optimism & enthusiasm • flexibility • confidence • empathy • patience • assertiveness • tolerance for change

The Danish library directors demand similar recruitment profiles and the new profile of their future staff will focus on personal skills, such as the ability to advise, to structure, to communicate, to teach and to market library services.

Since library managers are not in a position to replace all existing staff with new staff, we have to ensure access to the training and development that underpins the profile of a librarian today, and focus on personal skills and attitudes in addition to updating professional skills and competencies.

The distinction between knowledge, skills and attitudes is illustrated by an example from a workshop organised by the IFLA Section on Continuing Professional Development and Workplace Learning (CPDWL) held in Buenos Aires in August 2004 on Skills for Information Literacy Instruction. The participants in this workshop identified at least 35 different skills and techniques, which they considered relevant for the library staff involved in information literacy training. No order of priority was given. It might seem overwhelming for those librarians who are going to teach next week to meet all these suggested competencies. In the following section, I have tried to use the three categories (knowledge, skills and attitudes) to classify all the suggestions from the workshop - this exercise elucidated where formal continuing education is required and which new skills and attitudes can be acquired by training courses and in-house training. The need for knowledge as a backbone for information literacy training expressed in the workshop shows that librarians in this context are working on new ground and want competencies in topics, which are traditionally a part of schoolteachers' curricula.

Knowledge included foundations in areas such as: Philosophy of teaching.

- information literacy concepts;
- teaching styles and learning styles, learner-centred teaching;

- principles (active learning, case study methods, collaborative learning, peer critique and feedback, problem-based learning);
- evaluation (formative and summative methods).[1]

Skills included practical areas such as: Planning instruction (goals and objectives - lesson plan and outlines).

- presentation skills;
- asking and answering questions;
- managing time;
- discussion management;
- starting the session - handling questions;
- tools: handouts - presentation software -web pages - learning management systems;
- evaluation: evaluating your teaching, assessing student learning, evaluating a program.[2]

Personal Skills and Social Competencies are a combination of skills, which can easily be obtained by training courses, and skills and attitudes that are more linked to individual personality. These include:

- potential partners (getting in contact - agreeing on collaboration);
- connecting with students;
- voice and body language;

- creating a learning environment;
- resolving difficult situations.[3]

In RSLIS we cover a good part of the above-mentioned themes. A semester in our MLIS programme deals with the theoretical foundation for teaching information literacy, which today is also included in the RSLIS BSc. curriculum and assumes that teaching and instruction skills will be a natural part of the librarian's job. Training courses in themes mentioned under skills have been set up and taught with assistance from external teachers with pedagogical background, actors and journalists.

Development of new skills and competencies in local institutions

There are still some important aspects to ponder. How does staff maintain competencies that may have been acquired a long time ago? In many Danish libraries, some staff has been employed for more than 25 years and the average age is 50+. Also, how do we add new competencies at the speed demanded by the users' preferences and behaviour?

Library managers have to be forward looking and concentrate actively on competence development. It is not enough to dispatch the staff to yet another continuing education course or a conference. In order to undertake successful staff performance planning, it is necessary to start with an analysis of future tasks and roles of the research library and go through the four steps outlined in the figure below.

Skills identification, performance planning

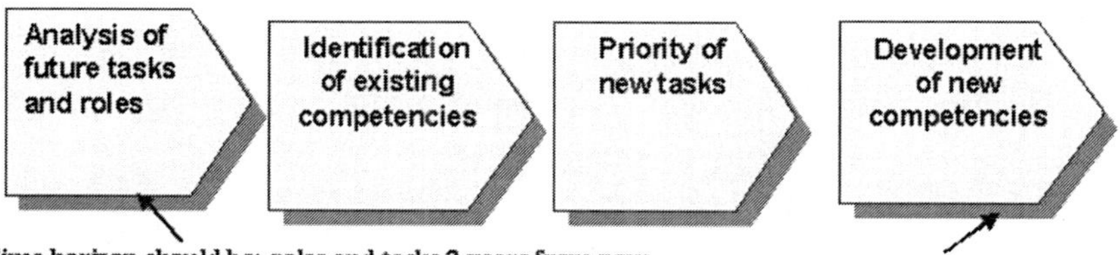

Time horizon should be: roles and tasks 3 years from now

The outcome of participation in formal continuing education is of course quite simple to measure - a new degree. The outcome of training courses, conferences and seminars might be of very high value for the individual.[4] For the entire organisation, however, it depends on the procedures for follow up and knowledge sharing - and many library directors still have a job to do in this area. Training courses organised locally for the entire staff or a department are shown to have a measurable impact and stimulate new activities and changes of attitude.

In Danish libraries, a variety of different methods are used for continuing professional development and training. We have, unfortunately, no research looking into which methods provide the most efficient results for training library staff.

From formal education and training to workplace learning

Broadening the concept of continuing professional education to that of continuing professional development (CPD) and workplace learning (WL) has been acknowledged by IFLA in accepting the new section CPDWL in 2003. This concept takes in all aspects of a professional's role, and recognises that education is only one of the ways in which individuals continue to learn and develop as professionals throughout their working careers and their professional lives. The concept of 'workplace learning' signifies the inclusion of different kinds of development activities, both the formal and less formalised, and includes such activities as work-based mentoring[5] and coaching. Beckett and Hager provide a detailed introduction to the concept and practice of workplace learning (Beckett & Hager, 2002).

My suggestions of methods for continuing professional development below are not in priority order or exclusive.

Suggested external activities for continuing professional development:

- Participation in projects across institutions/ across sectors.
- Participation in professional networks /interest groups (personal and virtual).
- Participation in formal continuing education programmes.
- Participation in external training courses (including distance education).
- Participation in conferences and seminars.
- Study visits.
- Job exchange.
- Preparing a talk for professional meetings.
- Writing a paper for publication.
- Preparing and teaching a course for library professionals.
- Establishment of new formal partnerships.

Suggested in-house activities for continuing professional development:

- In-house training / instruction by colleagues:
- Knowledge exchange with colleagues.
- Action learning / learning by doing.
- Job rotation.
- Reading professional literature.
- Testing new products / services.
- Mentoring programmes.
- Supervision.
- Benchmarking.
- Trial and error.
- Good management.

Resources needed

When planning for new skills and competencies it is necessary to take into account that sufficient resources should be made available within the institution. Participation in formal continuing education programmes, training courses and professional conferences requires separate budget allocations. Participation in external networks for knowledge exchange and conducting projects with colleagues from other libraries and across sectors seems to be very valuable for developing new competencies. Well-known in-house initiatives as listed above will also be usual for achieving new skills.

In the IFLA Section on Continuing Professional Development and Workplace Learning (CPDWL) there is an ongoing work on developing a set of recommended guidelines for continuing education. In this context we have been discussing the difficult issue: What is the minimum of resources needed for continuing education?

It is probably not possible to give a definitive answer to this question. I have asked a sample of Danish research library directors about their average resource allocation for continuing education and training. On this basis, I recommend the following figures: 10% of staff working hours as a minimum for training and professional development (external + internal activities). In terms of budget, 25-30% of the total budget should be allocated as a minimum for the development of new services. When considering the validity of this estimate, it is important to keep in mind that the figures reflect the state of development of Danish libraries - with decades of library automation experience, a well-functioning national coordination strategy and strong state support for library development - and not least library staff in all kind of libraries with a professional LIS education and continuing education programmes of a very high standard.

The seven circles of innovation – An innovation Management Model, 2005

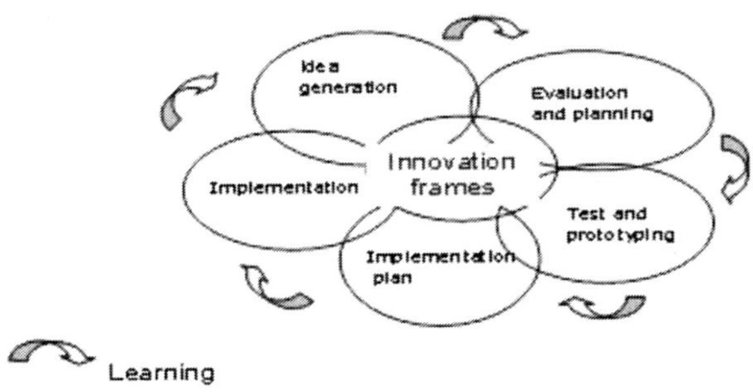

Three categories which reflect key dimensions in an organisation's innovation capacity		
Innovation frames - strategy, culture and competencies	**Innovation processes – from idea to implementation**	**Innovation results – "return on innovation"**
• Innovative attitude - customers - knowledge - analysis • Strategic focus - direction - communication • Culture and structure - attitudes - competencies - learning	• Management - support • Project management - steering - motivation • Work process - cooperation - clear goals	• Loans • Visitors • Quality • User satisfaction • Image

NOTES

1. New competencies by participation in formal continuing education, such as Masters Programmes in Learning and Teaching.

2. New competencies by participation in courses, professional interest groups, learning by doing.

3. New competencies by participation in courses, professional interest groups, learning by doing.

4. Methods for measuring the effects (outcome) of training new skills and competences (including the 'personal and social skills') exist, but are not frequently used by Danish libraries

5. For an example on work-based mentoring in the Karolinska Institutets Bibliotek, see: Haglund, 2004.

REFERENCES

1. Beckett, D. and P. Hager: Life and Learning: Practice in postmodernity. London: Routledge, 2002.

2. Bridges, Karl (Ed): Expectations of librarians in the 21st century. Westport, Conn.: Greenwood, 2003.

3. Corrall, Sheila: "Developing models of professional competence to enhance employability in the network world". In: Continuing Professional Development in Libraries: preparing for new roles in Libraries : a voyage of discovery : Sixth World Conference on Continuing Professional Development and Workplace Learning for the Library and Information Professions / ed. by Paul Genoni and Graham Walton München : Saur, 2005, 26- 40. [IFLA Publications ; 116]

4. Haglund, Lotta: "Mentoring as tool for staff development". Health Information and Libraries Journal. 21(2004), 61-65.

4

CHAPTER

Designing Libraries for Research Collaboration in the Network World: An Exploratory Study

Sheila Corrall

University of Pittsburgh, USA

ABSTRACT

Advances in technology and changes in higher education and research are forcing libraries to rethink the services they offer to researchers and how they are presented. Librarians are moving into "higher end support" and adopting new service models based on "deep collaboration" with academic partners. The internal arrangements that form the design of an organization can determine whether it achieves its mission under particular conditions. The present study explores how libraries in 24 leading UK research universities are organizing resources and services to support the research enterprise. Qualitative data were collected from institutional websites and other public domain sources and were analysed using matrix techniques. The results show a trend away from integrated library and computing service organizations; variety in institutional reporting lines, but predominantly in large professional service groups; consistency in internal library groupings, but variation in portfolio and job titles; expansion of specialist positions, with new functional roles complementing traditional subject liaisons; and dedicated spaces, working groups, and integrated websites promoting boundary-spanning activities. The findings confirm and extend prior work and are being used to design a large scale international survey.

Keywords: management structures, organization design, research support, university libraries

1. INTRODUCTION

Developments in digital media, network technologies, research workflows, scholarly communication and funding policies are challenging academic libraries to respond with a wider array of services and facilities for researchers (**Bourg, Coleman, & Erway, 2009**; **Housewright, Schonfeld, & Wulfson, 2013a**; **Vinopal & McCormick, 2013**). Subject/liaison librarians have traditionally provided research services as part of their academic support role, but libraries are increasingly identifying research support as a specific area requiring additional co-ordination and strategic development:

*"Open-access publishing, institutional repositories, the need to co-ordinate collection management and storage, the increasing availability of information technology (IT) tools to help researchers with their work and, not least, the Research Excellence Framework (REF) have all pointed to the need for a coherent library strategy and response" (**Bradbury & Weightman, 2010, p. 65**).*

University library support for research has emerged as a strategic issue and generated studies investigating service developments, new roles, skills requirements (**Auckland, 2012**; **Corrall, Kennan, & Afzal,**

2013; **Covert-Vail & Collard, 2012**; **Jaguszewski & Williams, 2013**). The present paper offers a different perspective on the topic by examining library engagement with research and researchers through the lens of organizational design.

1.1 Background

There is a perceived requirement for libraries to develop more specialized interventions to provide point-of-need research support and advice around the whole knowledge creation cycle: from ideas generation and project conception, through data acquisition, manipulation and interpretation, to the deposit of results, publication of findings, and assessment of impact (**Larsen & Riis, 2012**;**MacColl & Jubb, 2011**; **Vaughan et al., 2013**). Libraries are accordingly moving into areas such as funding opportunities and grant writing, ethics review, data curation and repository management, poster design and conference hosting (physical and virtual), journal and monograph publishing, bibliometric evaluation and impact assessment (**Adema & Schmidt, 2010**; **Carroll, 2011a**;**Crow et al., 2012**; **Delserone, Kelly, & Kempf, 2010**; **Drummond & Wartho, 2009**;**Furlough, 2010**; **Healy, 2010**; **Herther, 2009**; **Lyon, 2012**; **McBain, Culshaw, & Walkley Hall, 2013**; **Vaughan et al., 2013**), alongside their established roles in collection stewardship, resource procurement and information literacy. The development of support for the management of research data in line with national policy is a particular concern (**Carlson & Kneale, 2011**; **Gold, 2007**;**Lynch, 2008**; **Lyon, 2012**).

Commentators increasingly assert that libraries need to reposition from support service to professional/scholarly partner and to transform their liaison librarians into embedded experts recognized as academic associates (**Carlson & Kneale, 2011**; **Duranceau, 2008**; **Fonseca & Viator, 2009**; **Gold, 2007**; **Vandegrift & Varner, 2013**), so that they can add value through "higher end research support" (**Corrall et al., 2013, p. 638**). **Vaughan et al. (2013, p.**

313) argue that service presentation is an important factor here: "By basing the service model on the research lifecycle, the library is then poised to be a partner through the entire process, not just at the bookends of research". The proposed shift from the "service-and-support" approach to a partnership model is significant. **Posner (2013, p. 45)** argues that contemporary digital humanities projects "do not need*supporters* – they need *collaborators*", explaining that libraries need to provide both infrastructure (tools, servers, etc.) and "intellectual labor" (knowledgeable librarians). **Vandegrift and Varner (2013, pp. 69, 76)** similarly explain how partnering scholars in digital humanities should go beyond collection building to content creation, by "making 'stuff'" (e.g., web sites, digitized collections, new tools), describing the model as "deep collaboration".

Others have questioned whether libraries have sufficient expertise for such roles, particularly the technical understanding and domain knowledge needed for data management and curation in different disciplines (**Gold, 2007**; **Lynch, 2008**). However, feedback from biomedical and science librarians collaborating as partners in a large-scale cross-disciplinary multi- institutional research project reveals that personal attitudes and interpersonal skills are the key factors for successful participation in team science, with the willingness to learn and adapt more important than a deep understanding of the technology (**García-Milian et al., 2013**). Sector organizations including Research Libraries UK, the Joint Information Systems Committee, and the Association of Research Libraries (ARL), have sponsored projects examining the skillsets needed for new or extended subject liaison and specialist roles (**Auckland, 2012**; **Cox, Verbaan, & Sen, 2012**; **Jaguszewski & Williams, 2013**). Library and information schools are also responding with modules, courses, specializations, internships, and institutes to prepare new professionals and update the existing workforce (**Creamer, Morales, Kafel, Crespo, & Martin, 2012**; **Harris-Pierce & Liu, 2012**;**Lee & Tibbo, 2011**; **Stanton et al., 2011**; **Yakel, Conway, Hedstrom, & Wallace, 2011**).

The model of the librarian as a "scholar-practitioner" and active researcher is being promoted in several institutions (**Giesecke, 2010**; **McBain et al., 2013**;**Schrader, Shiri, & Williamson, 2012**), and being used to bring people with PhDs but no library degrees into research support roles in "alt/ac" (alternative to the academy) positions, particularly in relation to digital humanities developments (**Brunner, 2010**; **Giesecke, 2010**; **Nowviskie, 2013**; **Shore, 2012**). Even so, **MacColl and Jubb (2011, p. 10)** point out that despite recognizing their need for support, "researchers have little interest in the support services libraries have built for them". **Jahnke and Asher (2012, p. 16)** similarly report that "Few researchers see the library as a partner". Two large-scale surveys of academic faculty conducted in the US and UK confirm this message, with only half of the US respondents and one-third of the UK sample rating the library role in supporting their research productivity as "very important", placing this role as the least important of the six service areas specified (**Housewright et al., 2013a**; **Housewright, Schonfeld & Wulfson, 2013b**).

Some research libraries are responding to the challenge by moving beyond alignment with academic departments or schools to focus provision of in-depth information advice at the more specialized and dynamic level of research groups and projects, particularly in the biomedical and health sciences area (**Carlson & Kneale, 2011**; **Federer, 2013**; **Greyson & Surette, 2013**). Another significant trend is the creation of specialist positions dedicated to research support, either exclusively or in addition to a subject librarian role (**Bradbury & Weightman, 2010**; **Young & Lund, 2008**), including positions coordinating support for graduate students (**Covert-Vail & Collard, 2012**). The 2013 Ithaka survey of US library directors indicates that almost two-thirds of respondents in doctoral institutions expect to increase staffing for specialized faculty research support (**Long & Schonfeld, 2014**). Others have decided that traditional structures, such as the subject-specialist information consultant model, are no longer viable, and have reorganized their staffing to align with the broad institutional missions of research and teaching, or adopted business enterprise structures, with teams dedicated to service development and innovation (**Andrade & Zaghloul, 2010**;**Cotta-Schønberg, 2007**; **Covert-Vail & Collard, 2012**; **Drummond & Wartho, 2009**; **Franklin, 2009**; **Manchester University Library, 2012**; **Nutefall & Chadwell, 2012**).

1.2 Related work

Organization design has not received due attention in the debate on reshaping the library as research collaborator. Current literature on academic library structures consists largely of institutional case studies (**Andrade & Zaghloul, 2010**; **Chan & Soong, 2011**; **Franklin, 2009**; **Larsen & Riis, 2012**; **Nesdill, Love & Hunt, 2010**; **Quinlan & McHarg, 2012**; **Renfro & Neal, 2012**; **Shore, 2012**; **Warren, 2011**), with few comprehensive surveys. In North America, ARL last surveyed library organization in 1996, although the organization and staffing of specific elements of research support have been examined more recently, including digital humanities, research data management, and scholarly communication (**Bryson, Posner, St. Pierre, & Varner, 2011**; **Eustis & Kenney, 1996**; **Fearon, Gunia, Pralle, Lake & Sallans, 2013**; **Radom, Feltner-Reichert, & Stringer-Stanback, 2012**). In the UK, the convergence and merging of university libraries with information technology and other academic support services has dominated the discussion of library structures over the past two decades (**Bulpitt, 2012**; **Field, 2001**; **Hanson, 2005**; **Pugh, 1997a**).

Few writers on library organization relate their discussion to classic organizational theory, notable exceptions being **Pugh's (1997b, 2005, 2007**) efforts to establish a theoretical basis for the management of technology-driven change in libraries and **McGuigan's (2012)** essay identifying theories relevant to managing change in academic libraries. Although library case studies rarely use terminology from organization design literature, it is evident that the dominant library model of recent decades is basically the mixed structure (**Child, 1988**), which organizes technical/back-office processes (e.g., acquisitions, cataloging) and generic front-of-house services (e.g., circulation) along functional lines and specialized client services in a divisional or market-based arrangement (i.e., individuals or teams devoted to supporting particular subjects, disciplines, departments, or schools). Many libraries have overlaid an element of matrix structure, by introducing cross-cutting roles to coordinate service provision across different groups (e.g., designating senior liaison librarians as lead specialist for key areas, such as scholarly communications); an emerging variant of this model uses functional specialists who do not have liaison assignments to supplement and support their liaisons (**Jaguszewski & Williams, 2013**).

1.3 Definition of key terms

Organization design has been defined as "the allocation of resources and people to a specified mission or purpose and the structuring of these resources to achieve the mission" (**Duncan, 1979**, p. 60). **Duncan (1979, p. 60)** also asserts that "Ideally, the organization is designed to fit its *environment* and to provide the information and *coordination* needed" (emphasis added). **Stanford (2013, p. 19)** offers an alternative definition of the concept emphasizing other dimensions:

"Arranging how to do the work necessary to effectively and efficiently achieve a business purpose and strategy whilst delivering high-quality customer and employee experience. Arranging involves aligning the organization with the strategy, creating coherent designs, while building trust among key stakeholders."

Design is thus conceived as the broader higher-level concept, with *structure* forming an important part of the design. However, the terms are often used synonymously, within both management and library literature. Another important point to note here is that "resources" in this context can include "tangible resources such as money and equipment and intangible resources such as knowledge" (**Miles, Snow, Fjeldstad, Miles, & Lettl, 2010, p. 93**).

1.4 Research question, aims and objectives

The central research question for the study is: *How are academic research libraries and their institutions organizing resources and services to support the research enterprise?*

The overall aims of the research are:

- To review organizational designs of research libraries and evaluate their fitness for purpose in the context of researcher support needs.

The specific objectives of the project are:

- To review trends and developments in library support and services for research and researchers
- To examine design dimensions and structural features of a sample of research library organizations
- To identify specific elements of library organization of particular relevance in supporting research
- To analyze strengths and weaknesses of potential organizational models.

2. METHODOLOGY

This research uses an exploratory sequential mixed methods design (**Creswell & Plano Clark, 2011**) to survey organization structures of research libraries in the specific context of library service responses to challenges presented by the contemporary e-research environment. The purpose was first to qualitatively explore dimensions of library structural designs with a small single-country sample and then to quantitatively determine the extent to which the qualitative findings generalized to a large international sample. The present paper reports the results from the first phase only, but describes the complete design to explain the rationale for the initial research.

In the first (instrument-development) phase, data were collected via institutional websites, organization charts, strategy documents, job descriptions and other public-domain sources from a purposive sample of 24 universities in the UK. The Russell Group of leading UK research universities was used as the sampling frame.1 The use of websites and web-based documents as sources of data and objects of analysis has been recognized as an appropriate method of investigation for contemporary researchers, particularly in relation to organizational research (**Bryman, 2004**). Previous studies have used this type of Internet-based research to investigate library strategy documents, either as the sole method of data collection (**Corrall, 2007**; **Pacios, 2004**), or as the first phase of a multi-method study (**Aldrich, 2007**; **Corrall, 2008**; **McNicol, 2005**).

Both deductive and inductive strategies were used in the initial phase. A review of related literature informed the systematic browsing and keyword searching of websites, but the key outputs of the research (such as the categorizations of structural arrangements) were generated from the data analysis. Evidence was captured and recorded electronically as raw data (document extracts or whole documents), notes and initial codes. The data were analyzed using matrix analysis techniques, using color coding to highlight examples with striking features, which were then re-examined to select cases illustrating sharp contrasts (**Corrall, 2007**). Data matrices are particularly suited to making large amounts of qualitative data accessible and meaningful, enabling insights to emerge, and supporting both cross-site and within-site comparisons (**Miles & Huberman, 1994**; **Nadin & Cassell, 2004**).

The findings from this initial exploration will be used to develop assessment measures that can be administered to a large sample. In the planned second phase, data on service configurations, reporting relationships, divisions and specializations, coordination and integration will be collected from research library directors in Australia, New Zealand, North America, and selected European countries, using an online questionnaire survey. The reason for collecting qualitative data initially was that there is little guiding theory on the subject to inform the development of a survey instrument (**Creswell & Plano Clark, 2011, p. 86**).

3. RESULTS

Formal organization structure charts were found via web searches in eight cases, i.e., for one-third of the sample investigated (*n*=24). However, relevant dimensions and salient features of the structures in use at other sites were identified by browsing their web pages and consulting staff directories, service descriptions and other information available (including structure charts for the institution in some cases, and further particulars for job announcements in a few instances). The main elements examined were service configurations, reporting lines, internal organization, specialist positions, and coordination mechanisms. Throughout the analysis, the focus was specifically on elements of library organization of particular relevance in supporting research.

The headline findings from the data are as follows:

- Integrated library and computing services have fallen from favour
- Reporting lines for library and information services have shifted
- Library structures continue to mix function and market/subject elements
- New senior positions and titles are signalling strategic priorities
- Subject liaison arrangements are being enhanced and remodelled
- Specialist research positions are emerging as a significant trend
- Dedicated spaces for researchers are improving service visibility
- Groups and websites are promoting boundary-spanning services.

3.1 Service configurations

One-third of the services featured some form of convergence between the library and one or more cognate professional services, most commonly the computing/information technology (IT) service, which often also includes learning/education technology. This finding contrasts with the position a decade ago when more than half of the 20 institutions then in the Russell Group were converged and one-quarter had merged structures with fully integrated library and IT provision for several core functions (e.g., help desk, training). Among the currently converged organizations, four had retained the same basic structure for a decade or more, of which one was still a merged service, two were converged, and one was federated (separate organizations within an information services division of professional services).

Two organizations had deconverged their library and IT services and combined the library with other cognate services to form an academic services directorate and a student services directorate respectively. Two other institutions had formed newly converged services in the last five years, in both cases bringing the library and IT together, into an information directorate in one case, and academic services in the other (which also included student services and quality enhancement). However, the key finding here is that most members of the group now have separate library organizations, which includes four high-profile services that have deconverged their merged or converged service organization within the last three years (King's College London, the London School of Economics (LSE), and the Universities of Bristol and Nottingham).

The variety of service configurations was a feature of UK converged information service organizations in the 1990s and 2000s, and this pattern has continued as services have de-converged and re-converged more recently. For example, at both the University of Birmingham and the University of Exeter, Library services is one division within an Academic Services directorate of seven divisions, but the composition of each directorate and structure of the library division varies. Table 1 shows the breakdown (based on data collected in summer 2013).

At divisional level, in addition to library services, common functions here include academic development/educational enhancement, student employability (careers services), and student services/registry functions (combined at Exeter). However, technology services (including administrative systems, academic computing and student IT services) are part of the converged service at Exeter, but at Birmingham, in the re-converged service, IT Services is a separate unit within Professional Services reporting to the Registrar & Secretary – unlike the former converged information service organization at Birmingham, where IT was a central part of the original unified structure. At the library divisional level, Table 1 shows that the primary split into four functional sections is similar in both institutions, namely:

Table 1. Sample composition of academic services directorates

University of Birmingham		University of Exeter	
Academic Services Divisions	*Library Services Departments*	*Academic Services Divisions*	*Library & Culture Services Groups*
• Birmingham Foundation Academy • Centre for Learning & Academic Development and Learning Spaces • Student Employability • Library Services • Special Collections • Registry • Student Services	• Academic Liaison, Learning, Research Support & Academic Skills • Collection Management & Development • Library Customer Support	• Academic Support Unit (Cornwall) • Education Quality & Enhancement • Employability & Graduate Development • Exeter IT • Library & Culture Services • Planning & Resources • Student Services	*Library & Research Support Teams* • Academic Engagement • E-Strategy and Resource Development • Heritage Collections • Library Customer Services *Arts & Culture Team*

academic liaison/engagement; customer services/support; collection/resource development; and special/heritage collections. However, Birmingham has departed from established practice by making Special Collections a separate division (not reporting to the Director of Library Services), while Exeter has brought in Arts and Culture as an additional service group reporting to the Assistant Director of Academic Services/Head of Library and Culture Services.

The make-up of the academic support department/team within each library also differs slightly. The Birmingham department includes responsibility for the University's Academic Skills Centre (which provides support in areas such as academic writing, maths and stats, and study skills), in addition to academic liaison, and research support (including supporting the REF and research data management); whereas Academic Skills Development at Exeter is located within the Education Enhancement division, and the Library's Academic Engagement group includes separate teams specializing in digital content and library marketing, and open access and data curation, in addition to liaison support for teaching and research. (At Birmingham, responsibility for managing and supporting the institutional repository and open access is part of the Collection Development and Management department.) Both institutions use the traditional subject librarian model to carry out liaison activities: at the time of data collection, Birmingham had 15 Subject Advisors supporting 5 Colleges, while Exeter had 10 Subject/Liaison Librarians supporting 4 Colleges and separate Business and Medical Schools (part of the College structure at Birmingham). The numbers of liaisons are commensurate with the respective sizes of the populations served, Birmingham being approximately 50 percent larger than Exeter.

The Information Services directorates that have continued as merged, converged or federated services are now more uniform in their service portfolios. All include libraries (and archives), IT (academic computing and corporate systems), and learning technologies, usually including responsibility for both equipping and managing teaching spaces, but not covering academic skills (other than information literacy), which are typically provided by academic support/learning development units, nor other student services functions. Within the sample, although some instances of shared support functions were found (e.g., Operations and Central Projects at Birmingham; Strategic Support, Finance, HR, etc. at Cardiff; Planning and Resources at Exeter), only Edinburgh currently displays elements of the more radical merged service structures that were adopted to varying extents by Birmingham and other leading research universities from the mid 1990 s.

The key features of the Edinburgh structure are the combined library, e-learning and IT first-line help/support service, and the organization of liaison librarians and IT professionals into three College-facing consultancy teams within the User Services Division. The consultancy teams are in a different part of the organization from other library staff, who are mainly located in the Library and Collections Division. This means they are separated from staff working in related research and learning support services, including teams and sections dealing with scholarly communication, and collection development and management, and also do not report to the *de facto* library service head, the Director of Library & University Collections.

3.2 Reporting relationships

In UK universities, the Librarian traditionally reported to the Vice-Chancellor or Principal (as the chief executive officer of the institution), which often differentiated the library from the IT/computing service; but the adoption of more formal management arrangements in higher education institutions over the past two decades and the reconfiguration of administrative support functions into groups of professional services has resulted in different structures. Nearly three-quarters of the library directors in the sample report directly (or indirectly, via another senior officer) to the University's Registrar/Secretary/Chief Operating Officer. Chief Operating Officer (COO) is becoming the dominant title in use, found in half of the institutions.

Only Oxford and Cambridge seem to have retained their traditional reporting line in its original form, although the head of the converged service at Edinburgh (who holds the formal title of Vice-Principal for Knowledge Management, CIO and University Librarian) reports directly to the Principal. At the time of data collection, the university librarian/library director at Manchester and Imperial reported to the equivalent of the Deputy Vice-Chancellor (Deputy Rector at Imperial); at Leeds and Nottingham, they reported to a Pro-Vice Chancellor (PVC), in both cases the PVC with responsibility for education/teaching and learning. The grouping of the library with education/student services is another notable trend in structural arrangements, reflected in the recently converged academic service model adopted at Exeter, the reconverged academic services directorate at Birmingham, and reconverged student services at Queen Mary (University of London). In addition, the deconverged library service at King's now reports to the College Secretary via the Director of Students & Education; similarly at Bristol, the reporting line is via the Deputy Registrar, responsible for "the student-facing and education-related services".

It is debatable whether a PVC reporting line is preferable to the COO line, particularly where the PVC portfolio has a focus on teaching/students. Reporting directly to the COO role can position the library director alongside other professional service heads, including both student services and research services, which is particularly valuable in the context of collaborative service development with the research office or equivalent in areas such as managing research data and measuring research impact.

3.3 Internal organization

The size and shape of the library superstructures vary significantly, with leadership teams ranging from three to eight members. The titles of senior management positions and labels used for the main divisions/departments/operational units also vary, and some libraries combine one or more functions together, but overall there was a striking consistency in the way activities were grouped into the following five areas:

- Academic services/research and learning support (generally subdivided by subject field)
- Collection development/information resource management
- Special collections and archives/heritage collections
- Customer services/reader services/user support
- E-strategy/digital services/information technologies.

The academic (learning and research) liaison/support function was explicitly represented in the senior team in three-quarters of the sample. In one-third of these (six cases), the relevant manager had other responsibilities (e.g., Planning & Resources, Collection Development/Management, Special Collections, User Services, Service Development). In one novel example, at the University of Manchester, research services, teaching and learning support, and academic engagement were distributed among three different members of the leadership team, and combined with other related responsibilities. At Edinburgh, the merged Information Services organization separates consultancy including library liaison services (in the User Services Division) from other library-based research and learning services (in the Library & Collections Division).

Academic Services was the descriptor used most consistently, found in the job titles of seven senior managers. However Research, or Research & Learning (or Learning & Research) featured in seven titles of associate/assistant directors/heads of division or equivalent. The word Liaison was included in four titles, and the term Academic Engagement in two jobs. Other variants included the position of Head of Relationship Management at York, and the five Heads of Faculty Services in the senior team at Southampton (where academic liaison is directed by the Deputy Librarian). The most complex title identified was Assistant Director – Academic Liaison, Learning & Research Support, and Academic Skills at the University of Birmingham (a position that has subsequently been changed to Academic Engagement).

The collection development/management function was equally prominent, also being represented in the senior teams of three-quarters of the sample. However, the term "collection" was used explicitly in slightly less than half the sample, and the titles of the relevant roles, divisions, or sections also varied within these cases, e.g., Collection & Access Management (Nottingham), Collection Development & Description (Cambridge), Collection Services (Durham,

Leeds). The term “resources” seems to be gaining currency as an alternative to “collection”, suggesting the concept of the library collection may become less prominent as migration from purchased printed materials to licensed e-journals and e-books continues. At York, the relevant senior position is Head of Content.

In approximately half of the cases featuring a collection development or management role in the top team, the collections function was combined with the management or strategic development of library systems (in two cases along with facilities), represented by roles such as Head of Collections and eStrategy (Sheffield – a position that has now evolved into Assistant Director for Academic Services by adding responsibility for the Learning & Research Services team to the Procurement & Content and Digital Services & Systems teams), Head of eStrategy & Resource Development (Exeter – a position that is now combined with academic liaison responsibilities in an enlarged Academic Support portfolio), and Head of Information Management & Library Development (Bristol). A slightly less prevalent, but equally significant, trend (found in around one-quarter of the total sample), was the creation of senior roles solely concerned with the library’s information systems/information technologies portfolio, notable examples here including: Associate Director, Digital Library Programmes & Information Technologies (Oxford), Head of Digital Library Services (Leeds), and Head of Digital Technologies & Services (Manchester).

Around half of the institutions had special collections and/or archives represented as a distinct role in the library’s senior team at the level of divisional/departmental head, including two who used the alternative title of Head of Heritage Collections (Durham and Exeter). Also in around half of the sample, delivery of frontline services to users (such as lending/circulation and enquiry/help services) was represented explicitly in the senior group, e.g., Assistant Director – Customer Services (Kings), Head of Library Customer Services (Exeter). In several other cases, customer/user services was part of a larger portfolio for a member of the executive. Customer Services (or Customer Support) was by far the most common label used for this function, followed by four instances of User Services, two of Reader Services, and only one where the term Public Services was used.

3.4 Subject liaison

The importance of subject-based academic support was evident in the structures of all 24 libraries in the sample. With the exception of the Universities of Cambridge (where the library's strategic objectives include development of "integrated, subject-based provision for teaching and learning as well as research") and Manchester (where academic liaison through teams of faculty librarians was recently discontinued), every institution currently uses some form of subject librarian arrangement to provide specialized support to students and academic/research staff. Seven use the title *Subject* Librarian (or Subject Advisor or Subject Consultant) for this role, while 11 use *Liaison* Librarian (or Academic Liaison Librarian, or Faculty Liaison Librarian). (Other titles found include Academic Support Librarian, College Team/Support Librarian, Faculty Team Librarian, and Information Specialist.) In ten cases, the library had a team structure mirroring the institutional structure of Colleges or Faculties, with subject/liaison librarians led by Heads of Faculty Services, Faculty Team Leaders, Faculty Librarians, or similar. Some libraries had additional structural differentiation by including *Assistant* Liaison Librarians, or Liaison *Assistants* in their teams.

The strength of liaison teams/subject support varied across the sample from six liaison librarians at Queen Mary (University of London) and eight at Durham, to 32 subject librarians at Cardiff and 52 at Oxford (at the time of data collection). Although generally the liaison librarian title was associated with smaller-scale provision, this was not universally the case, as the Queen's (Belfast) structure included only eight Subject

Librarians (led by three Faculty Librarians), whereas Southampton had 24 Liaison Librarians, led by five Heads of Faculty Services (who also fulfilled liaison roles). The mean number across the sample was 15.9 and the median was 13.5. However, these figures should be seen as approximations only. They do not always represent the total number of librarians involved in subject-based learning support and research services, as many libraries have recently established specialist positions that focus solely on learning or research support, or a particular aspect of academic support. In addition, the method used to collect data on subject liaisons did not differentiate between full-time and part-time liaison staff.

3.5 Specialist positions

As reported in the Introduction, support for research has traditionally been a formal part of the subject/liaison/faculty librarian role, alongside support for teaching and learning, but published literature indicates a trend towards libraries putting more effort into the delivery and promotion of support for researchers, having identified opportunities to provide more specialized "higher-end" services, and also become aware that researchers often have a limited view of what librarians provide (tending to focus on resource procurement and collection management). A significant development in the past decade has thus been the creation of new roles within the academic services/support area designed to provide specialist support for researchers in areas such as literature searching, systematic reviews, current awareness, reference management, citation analysis, publishing advice, open access and data curation.

One-third of the sample libraries had established a specialist position with a research support focus, mostly within the past five years. In some examples, the role is similar to and positioned alongside existing subject/liaison librarians, with titles such as Academic Liaison Librarian (Researcher Support) at Durham, or Research Support Librarian at

Queen Mary. In most cases, the post-holder is also expected to fulfill a management coordination or strategic leadership role, both carrying out specialist work and providing direction for research-related service developments across library teams; examples of relevant job titles here include Research Support Leader at Liverpool, Library Liaison Manager (Research) at King's, Head of Scholarly Communications at Southampton, and Academic Support Manager (Research) at Warwick – where the post-holder is supported by a dedicated Academic Support Librarian. In a few instances, the post-holders also act as subject librarians/academic liaison contacts (e.g., Senior Consultant, Research Support at Cardiff, and Research Support Services Manager at LSE).

In addition to this type of cross-cutting coordination and development role, several libraries have established more specific research support positions, whose remits have evolved in line with additional needs identified. For example, the Research Skills & Development Librarian at Cambridge directs the University Library's Research Skills Programme, but is also a member of the University's Open Access Project team; at Imperial, the intended focus of the new position of Research Support Librarian (Medicine) was mediated literature searching for systematic reviews, but the role developed to provide support in other areas, such as data management. Other longer established examples of specialist research support provision in libraries include two largely grant-funded units, the Support Unit for Research Evidence (SURE, established in 2000) at Cardiff, which specializes in systematic reviews of the literature; and the Centre for Research Communications at Nottingham (established formally in 2009, but with a history going back to the SHERPA project of 2002–2006), which carries out research and development work in scholarly communication, open access and related areas.

Management of the institutional repository/research archive is now an established library function, but its location within the organization structure varies. In two instances within the sample, the repository function had been subsumed within the institutional research information system managed by the research office; in another two examples, the repository service was a joint venture with the university computing service, but located in the library. However, repository management was mostly a library responsibility, generally located either in the library's systems/IT/digital services division (10 examples), or with academic/research and learning services (9 examples). In two cases the research support librarian or equivalent was formally designated as the institutional repository manager, and in another two cases, the research support leader had formerly filled the role of repository manager. At Birmingham, the repository is the responsibility of the Collection Development and Management department, where the Digital Assets team is located (in the absence of a library IT department).

Policy developments and funding mandates related to research assessment, open access and data management have resulted in many new and redesigned specialist library positions and teams to manage and coordinate institutional support in these areas. For example, Edinburgh has a Scholarly Communications Team of four people providing support for policy development and implementation, open access advice and advocacy, intellectual property rights, bibliometrics, repository administration, publications submission, and metadata quality. Exeter has an Open Access and Data Curation Team, comprising a Manager, a Data Curation Officer, and an Open Access Administrator. At Bristol, an institutional Research Data Service was launched in summer 2013, as a Library-led collaboration with IT Services and Research & Enterprise Development; it is staffed by a Manager, two Senior Research Data Librarians and an Assistant Research Data Librarian. Sheffield has formed a new Research Support Team within its Learning & Research Services group, which includes two Coordinators for Open Access and Research Data Management, two Open Access Assistants and a Research Data Management Liaison Assistant.

3.6 Dedicated spaces

The University of Warwick Library launched its Wolfson Research Exchange in 2008 as an innovative suite of specialist services and dedicated technology-rich spaces for researchers, with provision for quiet study, interdisciplinary networking, academic meetings, and social activities, and a programme of events on topics such as funding opportunities, grant applications, data management, journal impact, and spinout companies, "offered not only by the library but by other university departments supporting research too" (**Carroll, 2011b, p. 55**). Novel services include Research Match ("a unique service which profiles your research and makes it visible") and Collaborative Sandpits (interactive multi-day workshops to create project proposals for funding).

With separate programming for graduate research students, early career researchers, and academics/research staff, the facility has its own dedicated staffing, with a Research Exchange and PG Hub Manager, a Coordinator, Support Officers, and Assistants, who work alongside other managers and academic support specialists in the Academic Services division of the Library. *In situ*support is provided by a team of part-time Advisers, who are all current or recent PhD students (**Carroll, 2011a**). The Library website presents the facility as "A space, a website, a community", reflecting its several dimensions, including its hosting of networking evenings, interdisciplinary special interest groups, online peer to peer discussions, and blogs.

The development of dedicated purpose-designed space for researchers emerges as a significant trend creating physical presence for renewed efforts by libraries to position themselves as research partners and collaborators. The new research spaces typically have card-controlled entry to restrict access and demonstrate that libraries understand that researchers want and need physical facilities that are separate from undergraduates. Queen Mary opened a new 58-seater Research Reading Room for graduate researchers and academic staff in 2011, after a consultation identified the need for a quiet, comfortable working environment away from busier noisier parts of the Library. The facility was "designed along similar lines to The British Library Reading Rooms, with high specification study desks finished in oak and leather, providing larger than standard worktop space, and LED task lighting at each desk". Birmingham, similarly, has a Researchers' Room "reserved for use by postgraduate research students and academics only".

Social space and technology provision generally differentiate contemporary researcher facilities from traditional reading rooms. For example, Durham is developing a researchers' study area for the exclusive use of postgraduate research students, with individual, collaborative and social areas, which is also being used to promote information about library support for research. At Exeter, the Old Library housing the University's special collections has been refurbished and renamed the Research Commons, with a new large reading room, enhanced IT, multi-media seminar room, and café-style break out space and terrace area. At York, where the Library was able to extend into an adjoining building, renovation has created new improved research space, including a study area, specifically for researchers; an informal research lounge area; and – the most unusual marketing feature – a Research Hotel, a suite of bookable rooms that visiting academic staff or researchers can reserve for their exclusive use as temporary work areas. The website explains that "The hotel concept does not extend to the provision of sleeping arrangements, but does offer use of the research lounge".

3.7 Boundary-Spanning Activities

Policy developments in the UK, notably the Engineering and Physical Sciences Research Council Policy Framework on Research Data (**EPSRC, 2011**) and the Research Councils UK (**RCUK, 2013**) Policy on Open Access have given rise to other structural arrangements, such as steering/working groups bringing institutional stakeholders together to facilitate compliance with funder requirements. Such groups are typically led or chaired by a Pro-Vice Chancellor for Research or senior research administrator and usually include representatives of academic schools/faculties/colleges in addition to key staff from the library, IT service and research office or equivalent. For example, at Cambridge, an Open Access Working Group was set up in 2012 to draft an institutional policy and make recommendations on implementation, followed by an Open Access Project Board in 2013 "to manage the transition to Open Access". At Exeter, "an Open Access and Research Data Management Policy Task and Finish Group was set up in March 2012 to develop an institutional policy on research data and Open Access issues". Leeds, University College London (UCL) and Warwick all have a two-tier structure for managing research data, with a high-level steering/oversight group or project board complemented by a Research Data Working Group at Leeds, a Technical & Operations Group at UCL, and an Action/Implementation Group at Warwick; the UCL structure also includes a User Group, which meets less frequently.

Institutional websites offer another opportunity to coordinate, focus and promote services to support research and researchers. Three-quarters of the library and information services in the sample had a set of web pages bringing together information targeted at researchers. At one-third of the sites, there was a prominent link from the library/information services home page, labeled Research Support, Researchers, or simply Research; at another eight, users could reach the relevant page with two clicks. The web pages varied in the topics featured, but there was a clear trend towards selectivity and integration, bringing together resources judged most useful to researchers, offered both by the library and other institutional units (such as research computing/IT services and the research office), and also trying to present the information from a research user viewpoint, rather than a service provider perspective. For example, at King's College London, the Library Research Support home page has the sub-title "Support through the research lifecycle" and a choice of six categories to click: Ideas, Research funding, Managing information, Disseminating research, Evaluating research, and Preserving research; there is also a link to the Library Research Support site from the institutional Research Support website. The Researcher@Library website at Leeds "gives researchers at the University of Leeds a single point of access to information, support and resources across the University".

Research impact, open access, and data management feature strongly on these web pages, reflecting current preoccupations of the UK research community, and also demonstrating how libraries are taking advantage of recent policy developments to position themselves as central to the research process. In almost all institutions in the sample, the library has been assigned responsibility for managing the funds allocated for Article Processing Charges incurred in compliance with funder requirements for gold open access. Most libraries have developed new web pages to support this process, typically including tables and/or decision trees/flow charts adapted to show how the institution is interpreting and implementing the relevant policies. Web pages supporting research data management were found at two-thirds of the institutions, with half of the sites clearly branded as part of the library/information services offer. Five services (including two projects led by the library) were presented as university-level offerings, usually acknowledging contributions by different services, and two services were based in the IT service. The websites were at various stages of development, with two consisting solely of links to external sites.

4. DISCUSSION AND CONCLUSIONS

4.1 Convergence

The move away from the converged service model is one of the striking findings of the study. However, the distinction between "organizational" and "operational" convergence (**Field, 2001**, p. 268) is important here. *Organizational*convergence (i.e., formally combining two or more services to form one organization) is evidently in decline within the sample, especially structural designs merging libraries with computing/IT services. *Operational* convergence (i.e., separate services/departments collaborating to coordinate their activities to improve conference and effectiveness), in contrast, is arguably more prevalent than ever, with libraries extending and deepening their collaborations and partnerships beyond IT and educational development colleagues to other professional services, such as research offices. This is particularly evident in:

- services (e.g., institutional repositories, research data services) delivered as collaborative ventures with IT services and/or research offices
- relationships and structures (e.g., committees and groups) supporting boundary-spanning activities in areas such as data management and open access
- unified research support websites providing a single point of access to university-wide resources for researchers, which could be described as an example of a *virtually* converged information service.

4.2 Positioning

The location of the library/information service organization in the institutional organizational structure is rarely discussed in the literature, but is important because it has implications for the allocation of resources, the coordination of activities, and the perception of the service by stakeholders. The two significant trends here were first, the administrative positioning of the library within professional services reporting to the Chief Operating Officer or equivalent, which was identified as the dominant model; and secondly, an increasing tendency to group libraries with "student-facing and education-related services" through reporting lines and directorate structures. The latter is arguably not a positive move for libraries trying to reposition themselves as credible collaborators and partners in research activities.

4.3 Specialization

The survey showed that departmentalization within libraries has evolved beyond the traditional reader/technical services model to a focus on five strategic areas of activity: information resources, academic engagement, customer service, heritage collections, and digital technologies, although there was significant variation in the size of leadership teams. Use of the term "Research" in the job titles of around one-third of the senior management positions identified sends a signal to stakeholders about the priority attached to this area. The strong focus on information systems/technologies identified at the senior management level is not surprising, but it was interesting to find that in many cases the technology portfolio was combined with another area of responsibility, and most frequently with information resources/collections, which in effect continued the tradition of locating systems librarians in a technical services division/department.

Specialization of library professional staff continues to focus on the subjects/academic disciplines taught and researched in the institution, but subject/liaison roles are increasingly being complemented and supplemented by coordinating roles (such as research support managers) and functional specialists for particular areas of activity (e.g., research data management, open access publishing). The hybrid approach to specialization found in UK research libraries echoes structural developments reported in North America (**Covert-Vail & Collard, 2012**; **Jaguszewski & Williams, 2013**).

An interesting dimension of specialization is the growing trend to extend the development of specialized research support into dedicated spaces for researchers. The space-as-service concept is gaining currency internationally as a visible contribution to research, particularly in support of early career researchers, and interdisciplinary collaboration, with the research commons beginning to emerge as a parallel development to the successful learning commons movement (**Corrall & Lester, 2013**; **Daniels, Darch, & de Jager, 2010**;**Ohio State University Libraries, 2012**). UK examples range from traditional reading rooms to the fully-serviced technology-rich space at Warwick, using early career researchers as peer support advisors, and offering novel services such as research profile matching and suites of bookable rooms for individuals and meetings (**Carroll, 2011a,b**).

4.4 Coordination

Coordination within libraries is being formalized through roles such as research support managers and leaders, and data management and open access coordinators (and also by internal committees and working groups). Coordination with other services and operations is being pursued through and facilitated by institution-wide groups at both strategic and operational level, typically established to deal with external mandates, but offering the potential of additional benefits for librarians in relationship building, professional credibility, and institutional positioning. As noted above, websites are an additional valuable way of coordinating research services to provide a single point of access for researchers and can be used not only to pull together library and other institutional resources and services, but again to position the library as a major player in the research arena. Research support websites complement the work of research support coordinators in libraries, but can also fulfill that role in libraries which do not have an individual tasked with that responsibility.

4.5 Outcomes

The purpose of the present study was to explore qualitatively how a sample of academic research libraries are organizing resources and services to support their research mission with the intention of identifying important dimensions of organizational design that could be used to develop an assessment instrument for a larger scale international survey. The study achieved its objectives and the results both confirm and extend the findings of related research investigating library plans, roles and skills for supporting and serving researchers. Although the sample was limited to 24 libraries in one country, the results identify structural variables related to library support for research that can be used as a provisional framework for libraries in the UK and other countries to examine and evaluate their own structures.

The present study was conceived as the first phase of a larger scale international survey of organizational design in research libraries and the findings are being used to develop a questionnaire to collect quantitative data on important structural and contextual variables identified from this exploratory study and ongoing review of related literature. The dimensions to be investigated include:

- positioning of libraries within their parent institutions
- scope of library/information service organizations
- size and focus of library leadership teams
- specialization of units, roles and space
- formal and informal convergence of activities
- mechanisms facilitating boundary-spanning activities.

NOTES

[1]http://www.russellgroup.ac.uk.

ACKNOWLEDGMENTS

A summary of this study was presented at the 2013 LIBER Annual Conference in Munich, Germany. A synopsis of the key findings was also published in *Library Connect Newsletter, 11*(3), The Library and the Research Enterprise (pp. 3–4).

REFERENCES

1. Adema, J., & Schmidt, B. (2010). From service providers to content producers: New opportunities for libraries in collaborative open access book publishing. New Review of Academic Librarianship, 16(sup. 1), 28–43.

2. Aldrich, A.W. (2007). Following the phosphorous trail of research library mission statements into present and future harbors. ACRL 13th National Conference, March 29–April 1, 2007, Baltimore, MD (pp. 304–316). Chicago, IL: Association of College & Research Libraries. Retrieved April 30, 2014

3. Andrade, R., & Zaghloul, R. (2010). Restructuring liaison librarian teams at the University of Arizona Libraries, 2007–2009. New Library World, 111(7/8), 273–286.

4. Auckland, M. (2012). Re-skilling for research: An investigation into the role and skills of subject and liaison librarians required to effectively support the evolving information needs of researchers. London: RLUK Research Libraries UK.

5. Bourg, C., Coleman, R., & Erway, R. (2009). Support for the research process: An academic library manifesto. Dublin, OH: OCLC Research.

6. Bradbury, K., & Weightman, A. (2010). Research support at Cardiff University Library. SCONUL Focus, 50, 65–70.

7. Brunner, M.L. (2010). PhD holders in the academic library: The CLIR Postdoctoral Fellowship Program. In K. Williams & S. Walter (Eds.), The expert library: Staffing, sustaining, and advancing the academic library in the 21st century (pp. 158–189). Chicago, IL: Association of College & Research Libraries.

8. Bryman, A. (2004). Social research methods. (2nd. ed.). Oxford: Oxford University Press

9. Bryson, T., Posner, M., St. Pierre, A., & Varner, S. (2011). Digital humanities: SPEC Kit 326. Washington, DC: Association of Research Libraries

10. Bulpitt, G. (Ed.) (2012). Leading the student experience: Super-convergence of organisation, structure and business processes. London: Leadership Foundation for Higher Education. Retrieved April 30, 2014, from http://www.lfhe.ac.uk/en/components/publication.cfm/S3%20-%2005

11. Carlson, J., & Kneale, R. (2011). Embedded librarianship in the research context: Navigating new waters. College & Research Libraries News, 72(3), 167–170.

12. Carroll, D.L. (2011a). Fostering a community of scholars at the University of Warwick: The Wolfson Research Exchange. New Review of Academic Librarianship, 17(1), 78–95.

13. Carroll, D. (2011b). What researchers want: A personal case study. SCONUL Focus, 51, 52–56.

14. Chan, D.L.H., & Soong, S.C. (2011). Strategic repositioning in a dynamic environment. Library Management, 32(1/2), 22–36.

15. Child, J. (1988). Organization: A guide to problems and practice (2nd ed.). London: Paul Chapman

16. Corrall, S. (2007). Benchmarking strategic engagement with information literacy in higher education: Towards a working model. Information Research, 12(4), paper 328.

17. Corrall, S. (2008). Information literacy strategy development in higher education: An exploratory study. International Journal of Information Management, 28(1), 26–37.

18. Corrall, S., & Lester, R. (2013). The researcher's view: Context is critical. In L. Watson (Ed.), Better library and learning spaces: Projects, trends and ideas (pp. 183–192). London: Facet

19. Corrall, S., Kennan, M.A., & Afzal, W. (2013). Bibliometrics and research data management services: Emerging trends in library support for research. Library Trends, 61(3), 636–674.

20. Cotta-Schønberg, M. (2007). The changing role of the subject specialist. Liber Quarterly, 17(3/4). Retrieved April 30, 2014

21. Covert-Vail, L., & Collard, S. (2012). New roles for new times: Research library services for graduate students. Washington, DC: Association of Research Libraries. Retrieved April 30, 2014, from http://www.arl.org/storage/documents/publications/nrnt-grad-roles-20dec12.pdf

22. Cox, A., Verbaan, E., & Sen, B. (2012). Upskilling liaison librarians for research data management. Ariadne, 70. Retrieved April 30, 2014, from http://www.ariadne.ac.uk/issue70/cox-et-al

23. Creamer, A.T., Morales, M.E., Kafel, D., Crespo, J., & Martin, E.R. (2012). A sample of research data curation and management courses. Journal of eScience Librarianship, 1(2), 88–96.

24. Creswell, J.W., & Plano Clark, V.L. (2011). Designing and conducting mixed methods research (2nd ed.). Thousand Oaks, CA: SAGE

25. Crow, R., Ivins, O., Mower, A., Nesdill, D, Newton, M., Speer, J., & Watkinson, C. (2012). Library publishing services: Strategies for success. Washington, DC: SPARC. Retrieved April 30, 2014, from http://wp.sparc.arl.org/lps/

26. Daniels, W., Darch, C., & de Jager, K. (2010). The research commons: a new creature in the library. Performance Measurement and Metrics, 11(2), 116–130.

27. Delserone, L.M., Kelly, J.A., & Kempf, J.L. (2010). Connecting researchers with funding opportunities: A joint effort of the Libraries and the University Research Office. Collaborative Librarianship, 2(1), 33–37.

28. Drummond, R., & Wartho, R. (2009). RIMS: The Research Impact Measurement Service at the University of New South Wales. Australian Academic & Research Libraries, 40(2), 76–87.

29. Duncan, R. (1979). What is the right organization structure? Decision tree analysis provides the answer. Organizational Dynamics, 7(3), 59–80

30. Duranceau, E.F. (2008). The “Wealth of Networks” and institutional repositories: MIT, DSpace, and the future of the scholarly commons. Library Trends, 57(2), 244–261.

31. EPSRC (2011). EPSRC policy framework on research data. Swindon: Engineering and Physical Sciences Research Council. Retrieved April 30, 2014, from http://www.epsrc.ac.uk /about/standards/researchdata/Pages/policyframework.aspx

32. Eustis, J.D., & Kenney, D.J. (1996). Library reorganization & restructuring: A SPEC kit. Washington, DC: Association of Research Libraries

33. Fearon, D., Gunia, B., Pralle, B.E., Lake, S., & Sallans, A.L. (2013). Research data management services: SPEC Kit 334. Washington, DC: Association of Research Libraries

34. Federer, L. (2013). The librarian as research informationist: A case study. Journal of the Medical Library Association, 101(4), 298–302.

35. Field, C.D. (2001). Theory and practice: Reflections on convergence in United Kingdom universities. LIBER Quarterly, 11(3), 267–289.

36. Franklin, B. (2009). Aligning library strategy and structure with the campus academic plan: A case study. Journal of Library Administration, 49(5), 495–505.

37. Fonseca, A.J., & Viator, V.P. (2009). Escaping the island of lost faculty: Collaboration as a means of visibility. Collaborative Librarianship, 1(3), 81–90. Retrieved April 30, 2014

38. Furlough, M.J. (2010). The publisher in the library. In S. Walter & K. Williams (Eds.). The expert library: Staffing, sustaining, and advancing the academic library in the 21st century (pp. 190–233). Chicago, IL: Association of College and Research Libraries. Retrieved June 23, 2014

39. García-Milian, R., Norton, H.F., Auten, B., Davis, V.I., Holmes, K.L., Johnson, M., & Tennant, M.R. (2013). Librarians as part of cross-disciplinary, multi-institutional team projects: Experiences from the VIVO collaboration. Science & Technology Libraries, 32(2), 160–175.

40. Giesecke, J. (2010). Finding the right metaphor: Restructuring, realigning, and repackaging today's research libraries. Journal of Library Administration, 51(1), 54–65.

41. Gold, A. (2007). Cyberinfrastructure, data, and libraries, Part 2. Libraries and the data challenge: Roles and actions for libraries. D-Lib Magazine, 13(9/10). Retrieved April 30, 2014

42. Greyson, D., & Surette, S. (2013). "You're just one of the group when you're embedded": report from a mixed-method investigation of the research-embedded health librarian experience. Journal of the Medical Library Association, 101(4), 287–297.

43. Hanson, T. (Ed.) (2005). Managing academic support services in university: The convergence experience. London: Facet

44. Harris-Pierce, R.L., & Liu, Y.Q. (2012). Is data curation education at library and information science schools in North America adequate? New Library World, 113(11/12), 598–613.

45. Healy, A.M. (2010). Increasing the visibility of the library within the academic research enterprise. Issues in Science and Technology Librarianship, 63.

46. Herther, N.K. (2009). Research evaluation and citation analysis: key issues and implications. The Electronic Library, 27(3), 361–375.

47. Housewright, R., Schonfeld, R.C., & Wulfson, K. (2013a). Ithaka S+R US faculty survey 2012. New York: Ithaka S+R. Retrieved April 30, 2014, from http://www.sr.ithaka.org/research-publications/us-faculty-survey-2012

48. Housewright, R., Schonfeld, R.C., & Wulfson, K. (2013b). Ithaka S+R/JISC/RLUK US survey of academics 2012. New York: Ithaka S+R. Retrieved April 30, 2014,

49. Jahnke, L.M., & Asher, A. (2012). The problem of data: Data management and curation practices among university researchers. In L. Jahnke, A. Asher & S.D.C. Keralis, The problem of data (pp. 3–31). Washington, DC: Council on Library and Information Resources.

50. Jaguszewski, J.M., & Williams, K. (2013). New roles for new times: Transforming liaison roles in research libraries. Washington, DC: Association of Research Libraries. Retrieved April 30, 2014

51. Larsen, S. & Riis, C. (2012). Aarhus University Library: Restructuring a university's library organisation. Bibliothek Forschung und Praxis, 36(2), 235–240. 0030.pdf

52. Lee, C.A., & Tibbo, H. (2011). Where's the archivist in digital curation: Exploring the possibilities through a matrix of knowledge and skills. Archivaria, 72, 123–168.

53. Long, M.P., & Schonfeld, R.C. (2014). Ithaka S+R Library Survey 2013. New York: Ithaka.

54. Lynch, C. (2008). The institutional challenges of cyberinfrastructure and e-research. EDUCAUSE Review, 43(6), 74–88.

55. Lyon, L. (2012). The informatics transform: Re-engineering libraries for the data decade. International Journal of Digital Curation, 7(1), 126–138.

56. MacColl, J., & Jubb, M. (2011). Supporting research: Environments, administration and libraries. Dublin, OH: OCLC Research.

57. Manchester University Library (2012). News from member libraries. SCONUL Focus,56, 61–62.

58. McBain, I., Culshaw, H., & Walkley Hall, L. (2013). Establishing a culture of research practice in an academic library: an Australian case study. Library Management, 34(6/7), 448–461.

59. McGuigan, G.S. (2012). Addressing change in academic libraries: A review of classical organizational theory and implications for academic libraries. Library Philosophy and Practice, Paper 755.

60. McNicol, S. (2005). The challenges of strategic planning in academic libraries. New Library World, 106(11/12), 496–509.

61. Miles, M.B. & Huberman, A.M. (1994). Qualitative data analysis: An expanded sourcebook. (2nd. ed.). Thousand Oaks, CA: SAGE

62. Miles, R.E., Snow, C.C., Fjeldstad, O.D., Miles, G., & Lettl, C. (2010). Designing organizations to meet 21st-century opportunities and challenges. Organizational Dynamics, 39(2), 93–103.

63. Nadin, S. & Cassell, C. (2004). Using data matrices. In C. Cassell & G. Symon (Eds.), Essential guide to qualitative methods in organizational research. (pp. 271–287). London: SAGE.

64. Nesdill, D., Love, A., & Hunt, M. (2010). From subject selectors to college and interdisciplinary teams. Science & Technology Libraries, 29(4), 307–314.

65. Nowviskie, B. (2013). Skunks in the library: A path to production for scholarly R&D. Journal of Library Administration, 53(1), 53–66.

66. Nutefall, J.E., & Chadwell, F.A. (2012). Preparing for the 21st century: Academic library realignment. New Library World, 113(3/4), 162–173.

67. Ohio State University Libraries (2012). Overview of research commons in academic libraries: A white paper. Columbus, OH: Ohio State University Libraries. Retrieved April 30, 2014,

68. Pacios, A.R. (2004). Strategic plans and long-range plans: Is there a difference? Library Management, 25(6/7), 259–269.

69. Posner, M. (2013). No half measures: Overcoming common challenges to doing digital humanities in the library. Journal of Library Administration, 53(1), 43–52.

70. Pugh, L. (1997a). Convergence in academic support services. London: British Library Research and Innovation Centre

71. Pugh, L. (1997b). Some theoretical bases of convergence. New Review of Academic Librarianship, 3(1), 49–66.

72. Pugh, L. (2005). The management of hybrid libraries. Library and Information Research, 29(92), 13–31. Retrieved April 30, 2014,

73. Pugh, L. (2007). Change management in information services (2nd ed.). Abingdon: Ashgate

74. Quinlan, C., & McHarg, H. (2012). The emerging library: Structure, culture, and lessons learned from the dissolution of a combined libraries–IT organization. Journal of Library Administration, 52(2), 147–161.

75. Radom, R., Feltner-Reichert, M., & Stringer-Stanback, K. (2012). Organization of scholarly communication services: SPEC Kit 332. Washington, DC: Association of Research Libraries

76. RCUK (2013). RCUK policy on open access and supporting guidance. Swindon: Research Councils UK. Retrieved April 30, 2014

77. Renfro, P., & Neal, J.G. (2012). The integration of libraries and academic computing at Columbia: New opportunities for internal and external collaboration. Journal of Library Administration, 52(2), 162–171.

78. Schrader, A., Shiri, A., & Williamson, V. (2012). Assessment of the research learning needs of the University of Saskatchewan librarians: A case study. College & Research Libraries, 73(2), 147–163.

79. Shore, E. (2012). Embracing hybridity: The merged organization, Alt/ac and higher education. Journal of Library Administration, 52(2), 189–202.

80. Stanford, N. (2013). Organization design: Engaging with change (2nd ed.). Abingdon: Routledge

81. Stanton, J.M., Kim, Y., Oakleaf, M., Lankes, R.D., Gandel, P., Cogburn, D., & Liddy, E.D. (2011). Education for eScience professionals: Job analysis, curriculum guidance, and program considerations. Journal of Education for Library and Information Science, 52(2), 79–94

82. Vandegrift, M., & Varner, S. (2013). Evolving in common: Creating mutually supportive relationships between libraries and the digital humanities. Journal of Library Administration, 53(1), 67–78.

83. Vaughan, K.T.L., Hayes, B.E., Lerner, R.C., McElfresh, K.R., Pavlech, L., Romito, D., Morris, E.N. (2013). Development of the research lifecycle model for library services. Journal of the Medical Library Association, 101(4), 310–314.

84. Vinopal, J., & McCormick, M. (2013). Supporting digital scholarship in research libraries: Scalability and sustainability. Journal of Library Administration, 53(1), 27–42. Retrieved June 23, 2014

85. Warren, V. (2011). Using workforce structures to enable staff retention and development: An academic library case study. New Library World, 112(1–2), 8–18.

86. Yakel, E., Conway, P., Hedstrom, M., & Wallace, D. (2011). Digital curation for digital natives. Journal of Education for Library and Information Science, 52(1), 23–31. Retrieved June 23, 2014.

87. Young, H., & Lund, P. (2008). Reflections on a benchmarking survey of research support provided by 1994 Group libraries. SCONUL Focus, 43, 51–56.

5
CHAPTER

On the Role of Research Data Centres in the Management of Publication-Related Research Data

Sven Vlaeminck, Gert G. Wagner

German National Library of Economics/Leibniz Information Centre for Economics (ZBW), Hamburg

German Data Forum (RatSWD), German Institute for Economic Research (DIW Berlin), Max Planck Institute for Human Development, Berlin, and Berlin University of Technology (TUB)

ABSTRACT

This paper summarizes the findings of an analysis of scientific infrastructure service providers (mainly from Germany but also from other European countries). These service providers are evaluated with regard to their potential services for the management of publication-related research data in the field of social sciences, especially economics. For this purpose we conducted both desk research and an online survey of 46 research data centres (RDCs), library networks and public archives; almost 48% responded to our survey. We find that almost three-quarters of all respondents generally store externally generated research data – which also applies to publication-related data. Almost 75% of all respondents also store and host the code of computation or the syntax of statistical analyses. If self-compiled software components are used to generate research outputs, only 40%

of all respondents accept these software components for storing and hosting. Eight out of ten institutions also take specific action to ensure long-term data preservation. With regard to the documentation of stored and hosted research data, almost 70% of respondents claim to use the metadata schema of the Data Documentation Initiative (DDI); Dublin Core is used by 30 percent (multiple answers were permitted). Almost two-thirds also use persistent identifiers to facilitate citation of these datasets. Three in four also support researchers in creating metadata for their data. Application programming interfaces (APIs) for uploading or searching datasets currently are not yet implemented by any of the respondents. Least common is the use of semantic technologies like RDF.

Concluding, the paper discusses the outcome of our survey in relation to Research Data Centres (RDCs) and the roles and responsibilities of publication-related data archives for journals in the fields of social sciences.

Keywords: Research Data Management, Research Data Centres, Journals, Libraries Economics

1. BACKGROUND AND INTRODUCTION

In the social sciences (especially economics, political science and sociology) more and more researchers analyse data provided by official statistics or by specialised providers of research data (e.g., from the ALLBUS at GESIS1 or from the SOEP at DIW Berlin2). In addition, relevant data may often also be purchased from companies like Thomson Reuters or Bloomberg.

Especially in economics, compared to other branches of empirical research, the compilation of own datasets is not common. A major exception is the field of experimental economics, where researchers often generate their own datasets in the course of investigations motivated by game theory.

Although a rising number of publications in almost all scientific disciplines are based on the analysis of datasets, there are few effective ways to effectively replicate or re-examine the results of an empirical article, to verify it, or to make it available for re-utilisation and to support scholarly debates.

Even research data, which — in principle — is publicly available, will not typically be archived (e.g., in a final working-file) with respect to the specific selection and adjustment procedures. Therefore, while replications are not necessarily prevented, they are extremely difficult in the cases of ambitious analysis based on specific data selections and calculations.

The current situation confronts both the scientific community and scientific infrastructure service providers, like libraries and research data centres, with multiple challenges. In addition to questions concerning data availability and incentives for sharing data, there exist also infrastructure challenges. In particular the roles and responsibilities of scientific infrastructure providers, e.g., research data centres (RDCs), for managing and operating a data archive that facilitates the replications of published research often are not clearly outlined.

The first part of our paper describes some of the problems that lead to poor replicability of social sciences research. Then our paper describes the outcome of desktop research and an online survey evaluating scientific infrastructure with respect to their potential services for the management of publication-related research data in the field of social sciences.

The conclusion of our paper discusses the roles and responsibilities of several stakeholders for operating data archives for scholarly journals. Experiences in other scientific areas are integrated in our suggestions for establishing data archives that are based on the complementary know-how of research data centres (RDCs) and libraries.

1.1. Why is social science research often not replicable?

According to the literature the following reasons for missing replicability may be mentioned:

- First, and most importantly, there is a lack of incentives for researchers to share their data with the community. The academic reward system does not honour the often time-consuming efforts of data sharing — in sharp contrast to publications, although *"[a]n applied economics article is only the advertising for the data and code that produced the published results"* (**Anderson, Greene, McCullough, & Vinod, 2008**, p. 101).

- Furthermore, social scientists may worry that data sharing could lead to personal disadvantages. Researchers who work up and share data with the community do not receive appropriate compensation, e.g., reputation, for their efforts and may even suffer in terms of academic career because data sharing takes time that cannot be spent on research. In addition, many researchers suspect others will "misuse" their data, for example with faulty interpretations or by using a dataset without due reference to the creator of the dataset (**Fecher, 2014**). Eventually, the legal status of research data with regard to data sharing is not sufficiently clear, which also leads to reservations in data sharing.3

- Few social science journals have currently implemented guidelines requiring their authors to provide their data and statistical computation codes (**McCullough, 2009; Vlaeminck, 2013**). So called "data availability policies" may, in some instances, oblige the authors of empirical research papers to supply the underlying data of their results and the code/syntax of their analysis along with the manuscript of the article. Those policies often are in line with the "replication standard" formulated by Gary **King (1995)**.

- Useful infrastructure components for the management of publication-related research data are rarely applied, which, in turn, prevents any uniform way of citing the underlying data. Available technical solutions like Dataverse,4 a powerful tool for managing and documenting publication-related research data, are adopted by only a few journals. In this context a critical point focusses on how professional research data centres handle research-related data and what kind of services, if any, they offer.

1.2. Do research data centres offer services for archiving publication-related research data?

Research data centres could actually be ideal institutions for managing publication-related research data published as attachments to articles within scholarly journals. These capacities originate from decades of expertise in the handling of social- and economic research data, from core-competencies in the creation and maintenance of metadata collected and tagged from surveys as well as extensive experiences in managing access to data (**Research Information Network, 2011**). **Cox and Pinfield (2013)** argue that librarians, in contrast, already often feel over-taxed with the multiple roles that they have in the various activities of their libraries. In addition, libraries may lack technical knowledge, domain-specific expertise and may also have limited personal experience in the common research processes. As such it may be difficult to position libraries as key players in this area. A loophole could be the close collaboration between libraries and research data centres to solve upcoming challenges, as **Christensen-Dalsgaard (2012)** suggests.

Therefore, the EDaWaX (European Data Watch Extended – www.edawax.de) project, funded by the German Research Foundation (DFG) 2011 to 2016, conducted a study evaluating if such services for publication-related research data are currently available from scientific infrastructure service providers like research data centres, libraries and archives. For this purpose a list of 46 scientific infrastructure organisations was prepared. It includes all German research data centres and data service centres accredited by the German Data Forum (RatSWD),5 research data centres organised within the Council of European Social Science Data Archives (CESSDA),6 the library networks in Germany as well as individual libraries and public archives.

Our investigation into the services provided by these data centres for managing publication-related research data in the social sciences is a hands-on approach to evaluate the possibility of cooperation between research libraries and research data centres. Therefore our study followed the suggestion of **Lyon (2012)** to develop "*a proactive approach to collaborating with disciplinary, national and international data centres ... for data deposit in such archives*" (p. 130).

In a first step, the websites of these organisations were examined with regard to potential services for storing and hosting publication-related research data. The ICPRS (Inter-university Consortium for Political and Social Research — University of Michigan) provides a publication-related archive7 that is used by numerous authors to deposit their publication-related data.8 NARCIS,9 a research information system located in the Netherlands, offers a specific service for publication-related data.10 DANS EASY,11 another service located in the Netherlands, can also be used to deposit such data in principle.12

However, desk research could not uncover other information needed for further analysis, which is why, in order to start a more detailed evaluation of potential services by these organisations, an online survey was conducted.

2. THE ONLINE-SURVEY

In October and November 2012 an online-questionnaire was sent to 46 organisations — among them 35 national and international research data and data service centres, 1 archive, 7 library networks and single libraries, as well as three other organisations (non-European research data centres). A satisfactory, especially when compared to average return rates of mail survey.

Due to the structure of the questionnaire, not all participating organisations responded to all questions, which explains deviations in the number of responses (Figure 1).

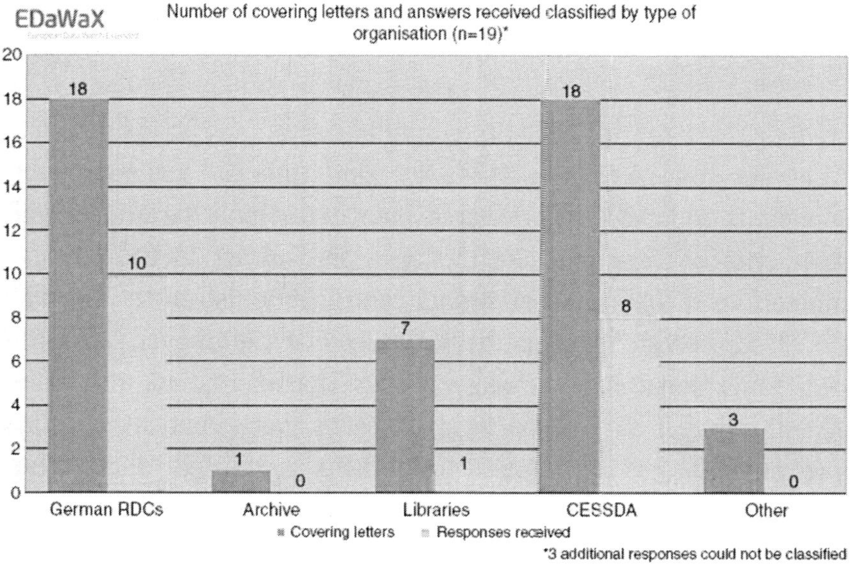

Figure 1. Covering letters and responses received.

Certainly more important than the return rate is the structure of respondents and non-respondents. The large majority of responses came from research data centres in Germany and Europe (86%). Significantly under-represented were respondents from German library networks and archives. The three non-European research data centres did not respond.

We can only presume that the library networks and the archive do not offer relevant services for research data management and, therefore, did not respond to our survey.

2.1. Empirical findings

Initially, the survey asked whether institutions would, in principle, host and store publication-related research data.13 In addition, the survey also asked whether organisations would also host and store (self-compiled) software components and the code of computation/syntax of statistical analyses. These three types of data are often part of empirical submissions to economic journals.14

2.2. Datasets

More than three-fourths of all organisations responding accept external datasets for storage (Figure 2). At the same time the lion's share of respondents reported that research data would only be accepted if certain criteria were met. Such criteria are subject to the specific competencies of many research data centres, but also to the specific regional/supra-regional or national competencies. Moreover, technical and organisational aspects (e.g., proper documentation, machine-readability, etc...) as well as legal concerns were cited as criteria. Approximately 74% of the respondents indicated that their organisations would host these types of data (Figure 3). If any criteria for hosting were mentioned, the subject-specific orientation of an institution was stated as main criterion

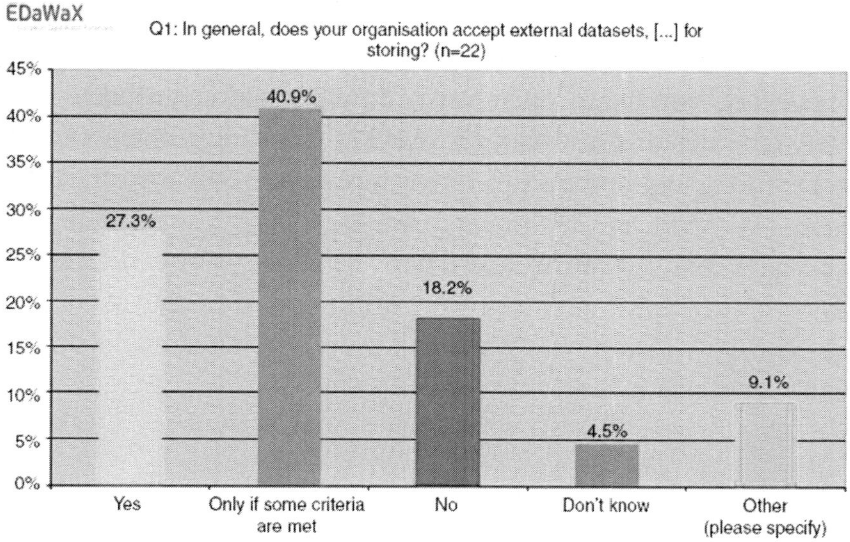

Figure 2. Acceptance of externally created datasets for storing.

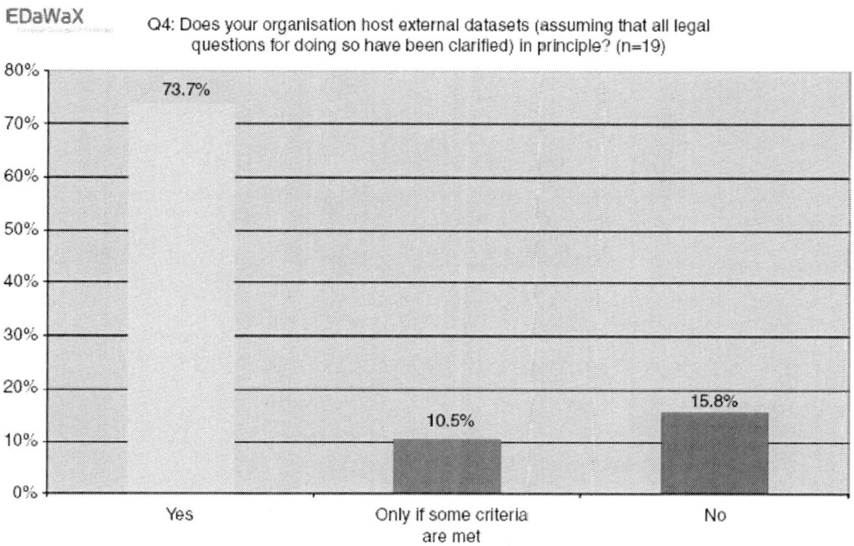

Figure 3. Hosting of externally created datasets.

2.3. Software

With regard to storing and hosting of (self-compiled) software components, which are often used for economic simulations, our survey indicates that just under a fourth of responding organisations accept storing and hosting software components without restrictions (Figure 4). Another 17% pointed out that they have criteria for assessing if software can be stored and hosted (e.g., *if essential for the analysis of the data*). Therefore, a gap exists in the availability of hosting and storing software components. Only a limited number of organisations offer this service.

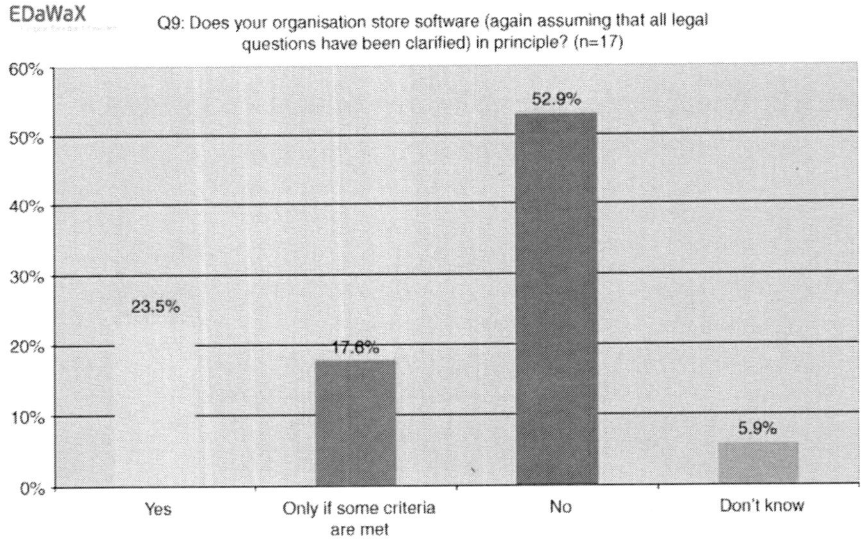

Figure 4. Storage of software components (e.g. used for simulations).

2.4. Code of computation/Syntax of statistical analyses

Almost 70% of the organisations responding offer options to store and host computation codes (Figure 5). However, a quarter does not do so at present and is not considering offering such services in the near future. One respondent also stated a criterion — noting that the storing and hosting of these data would only be useful in the case of derived variables.

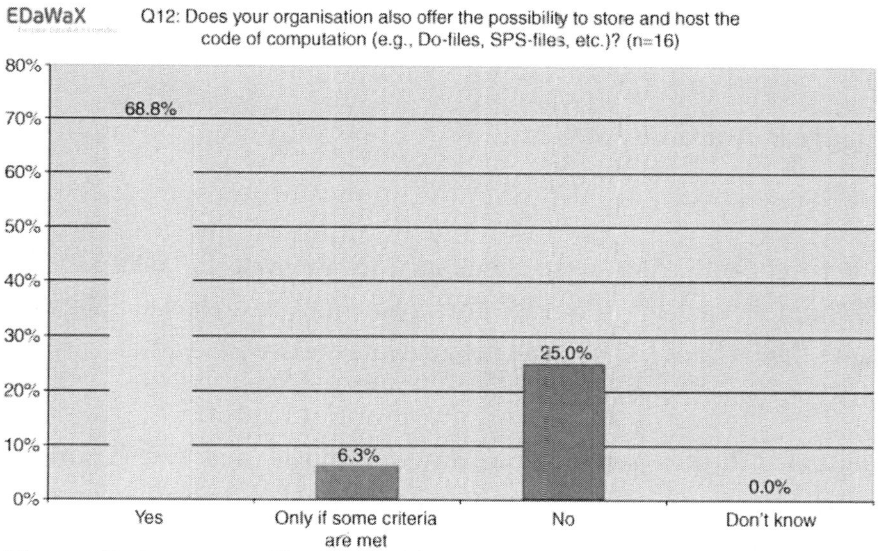

Figure 5. Storage and hosting of the syntax of statistical analyses.

2.5. APIs

Within our analyses we also examined the availability of application programming interfaces (APIs), which enable automated data exchanges. Our results show that less than half of all responding organisations have these interfaces at their disposal (Figure 6).

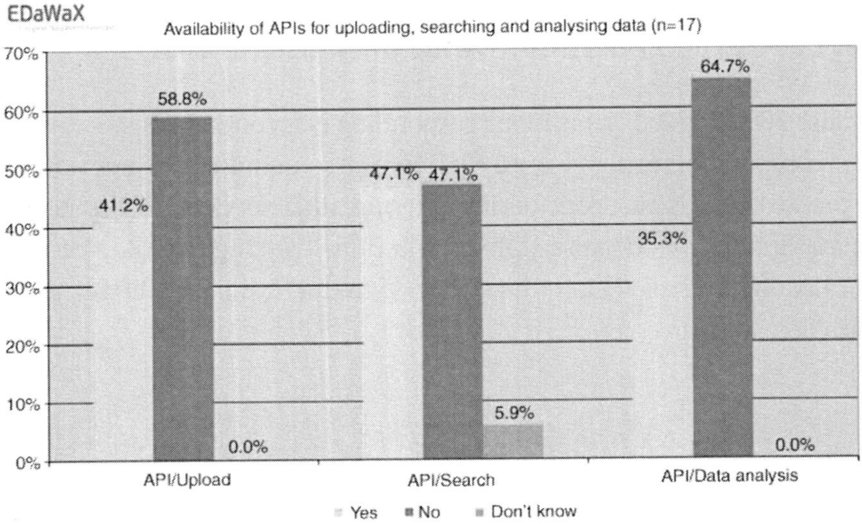

Figure 6. Availability of APIs.

Most frequently APIs were mentioned as a device for data search (47%), followed by APIs used for uploading research data. Slightly more than a third (35%) of all respondents declared the availability of an API to analyse research data.

However, further analysis by EDaWaX shows that the reported interface consists only of searching and uploading interfaces on the respondents' websites. We were not able to find an API. Presumably, APIs in terms of external reading and writing accesses are by and large unknown among our respondents and not readily available.

2.6. Metadata schemata and the creation of metadata

2.6.1. Employed metadata schemata

We were also interested in the metadata schemata currently used by the organisations in their daily work. Our survey shows that more than 70% of the respondents use DDI (Figure 7). Other schemata like Dublin

Core are rarely used (29%).15 All other metadata schemata are used rather sporadically.

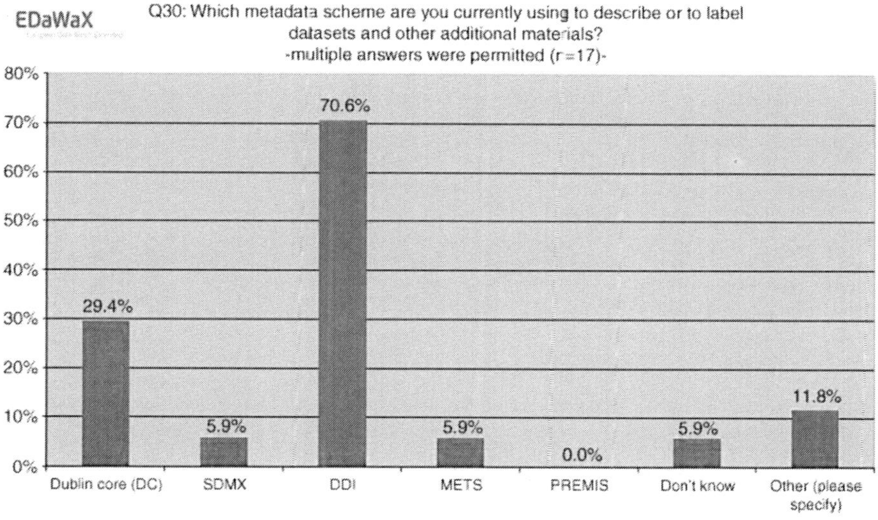

Figure 7. Metadata schemata currently in use.

2.6.2. Persistent Identifiers (PI)

In addition, we asked, whether organisations assign persistent identifiers (e.g., handle, DOI, URN, etc…) to datasets and other materials. The persistent identification of research data is an important issue, for instance because it enables researchers to cite datasets.

More than 56% of the organisations in our sample assign such identifiers by default, but almost a third do not (Figure 8). The persistent identification of research data is an important issue, for instance because it enables researchers to cite datasets.

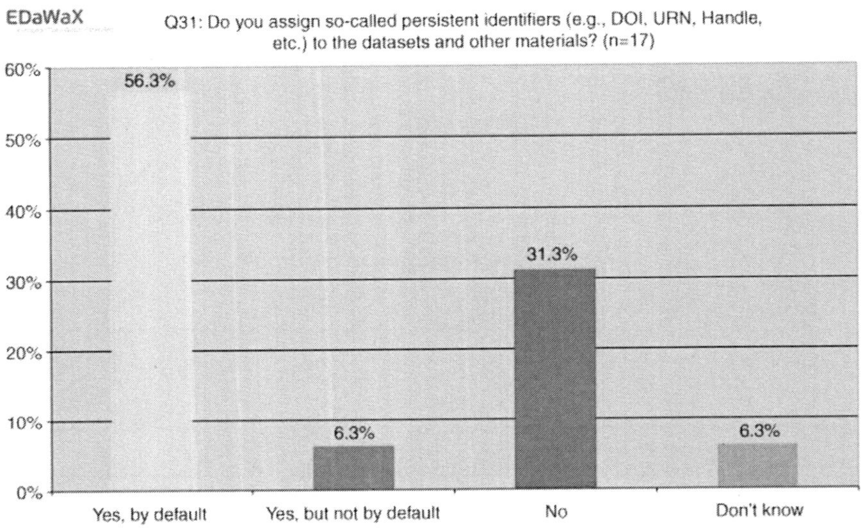

Figure 8. Assignment of persistent identifiers.

2.6.3. Support of Semantic Web Technologies

In our survey we also examined the implementation of RDF (Resource Description Framework). RDF is a general method for conceptual description or modelling of information implemented in web resources. Among the organisations answering this question a minority of 6% claimed to use and disseminate RDF-files. Almost a quarter of all respondents did not specify whether their organisation uses RDF, which presumably indicates that RDF is largely unknown.

2.6.4. Support for creating metadata

Again and again, a critical issue regarding the reuse of research data is the quality of data documentation. Therefore, a matter of particular interest is whether respondents support researchers in generating metadata and, if so, how.

Our survey shows, that the majority (almost 65%) of all responding organisations do so (Figure 9).

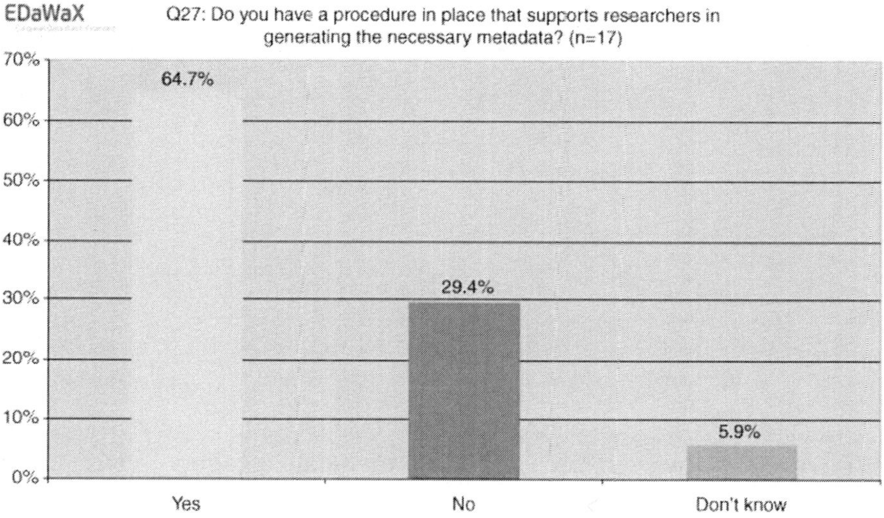

Figure 9. General support for metadata creation.

Furthermore, we were keen to know whether this support is software-based — e.g., if there is a web frontend where researchers may type in the required information that is then converted into a standardised metadata schema.

We find that 36% of the respondents use this type of software-based support with researchers (Figure 10).

There are a striking number of statements in the section *other*. Part of the *other*support for researchers, for instance, consists of written *data deposit forms*.

Our question regarding the software program names revealed that at least two institutions use Nesstar.16 Many organisations also use in-house solutions.

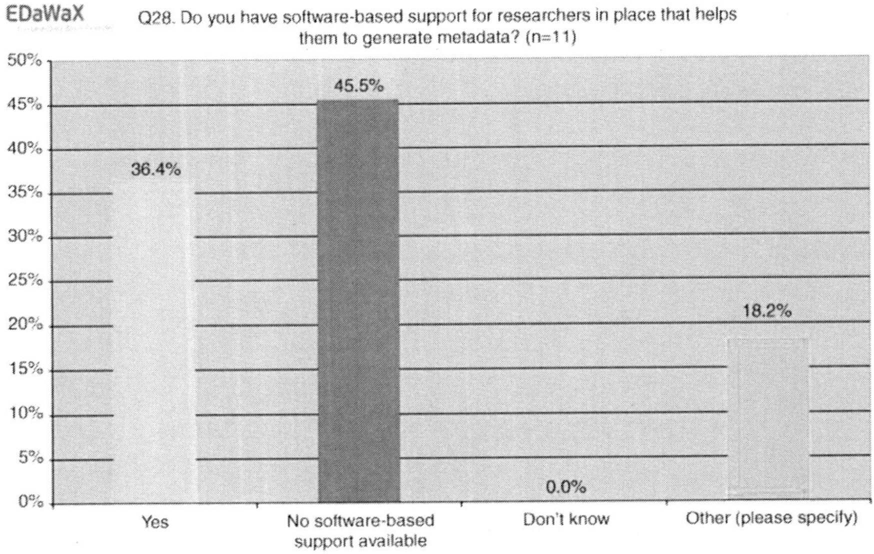

Figure 10. Software-based support for metadata creation.

2.6.5. Digital long-term preservation

In our survey we wanted to identify to what extent the respondents' institutions have implemented specific measures for long-term research data preservation. Therefore we asked the respondents whether their organisations take specific actions for digital long-term preservation. Because format-migration is one of the dominant strategies for long-term preservation (**Harvey, 2012**), we suggested format migrations as one such method.

Our survey indicates that more than 80% of all organisations use these types of procedures (Figure 11).

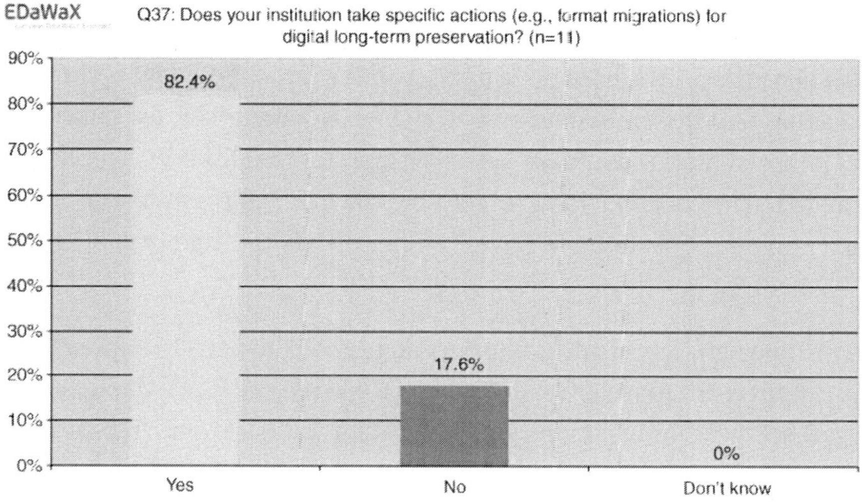

Figure 11. Long-term preservation of research data.

3. CONCLUSION AND DISCUSSION

Our study aims to evaluate if services for publication-related research data are available from data centres, libraries and archives. Based on existing services, our project defines roles and responsibilities for operating a publication-related data archive for journals in the fields of social sciences. This approach is in line with numerous recommendations put forth by European and national organisations as well as projects to interrelate research outputs with their underlying research data (**German Council of Science and Humanities, 2012**;**Kroes, 2012**; **Reilly, Schallier, Schrimpf, Smit, & Wilkinson, 2011**).

A question often arising in the context of linking data and publications is the discussion about stakeholders and their roles and responsibilities in the process (**Costas, Meijer, Zahedi, & Wouters, 2013**; **Lyon, 2007**). At first glance publishers appear to be the optimal stakeholders to perform the task of building up effective and efficient data archives because many publishers already host supplementary material for their journal articles. Therefore developing and implementing data archives, collecting and disseminating research data and metadata for datasets and other material could be an easy task for publishers. Therefore wouldn't it be a good idea to rely on the publishing industry? For answering the question we have to differentiate the role of academic publishers.

On the one hand, currently publishers do not see the need to implement data archives for journals on their own (**De Waard, 2012**). One reason might be that implementing and operating data archives raises the costs of publication. On the other hand, the availability of a data archive does not necessarily increase the number of journal subscriptions. Hence, the incentives to build up and operate data archives are not readily apparent to publishers.

In addition, questions of ownership and access conditions to archived research data could cause uncertainty for researchers, despite a publisher's announcement "*not to require any transfer of or ownership in such data or data sets as a condition of publication of the article in question*" (**STM & ALPSP, 2006**, p. 1).

Despite the fact that publishers do not operate data archives for their journals, they can nevertheless play an important role in the process of interrelating research data and publications. We already observe such collaborations in some scientific disciplines where publishers and data archives actively cooperate. In disciplines like the earth sciences, the e-infrastructure needed for storing and hosting research data in conjunction with appropriate documentations of the data has already been ongoing for quite some years. From a publisher's perspective linking research data and publications provides a benefit for their journals if the scientific outputs that are enriched with research data generate more citations (**Piwowar, Day, & Fridsma, 2007**). In addition, these links enable a more accurate research process and offer protection against scientific misconduct (**McCullough, 2009**).

Excellent examples of collaborations between publishers and data repositories include PANGAEA and Dryad. PANAGEA is the "data publisher" for earth and environmental sciences. It partners with Reed-Elsevier.17 Dryad is a non-profit repository for data underlying the international scientific and medical literature. It partners with numerous journals.18

Based on these experiences, the best solution is to implement and operate a discipline-specific data archive that gains importance by acquiring more and more data, which subsequently partners with publishers. The evolution and success of PANGAEA and Dryad underlines this approach impressively.

So, if it is not up to the publishers to run a data repository, other stakeholders come to the fore. In particular, research libraries and research data centres are the best positioned to take on the responsibilities of running such a disciplinary data repository. The results of our empirical investigation lead us to the conclusion that research data centres (RDCs) are likely the most relevant places to taking on the role of hosting and storing publication-related research data that is submitted to journals. RDCs already meet many prerequisites. In particular, the RDCs we analysed, in the broader field of social sciences, have much data handling experience. They are well trained in the storing, handling and documentation of these types of data as well as in taking appropriate measures for long-term data preservation.

Because RDCs currently do not comply with all requirements with respect to storing and hosting publication-related research data, collaborations between libraries and research data centres appear to be a promising way for establish such data archives: Libraries have the skills for managing publications. These include a dedicated knowledge of using authority files and multiple metadata schemata, in cataloguing information and providing this information to their discovery systems. Or as James L. Mullins, Dean of Purdue University's Library, describes it, "*Our ability to see structure to overlay on a mass of disparate 'parts,' as well as the ability to identify taxonomies to create a defined language for accessing and retrieving data is what is needed from us*" (**Baykoucheva, 2011**, p. 46). Unlike RDCs, it seems to be much more common for libraries to provide their stocks to their customers and to implement technical systems and the APIs necessary to do so.

According to **Pullinger and Wagner (2010)**, managing research data comprises of a mix of information that goes beyond the traditional separate realms of publications (the primary responsibility of national libraries), official records (the responsibility of national archives) and datasets (the responsibility of researchers themselves, statistical offices and RDCs) (**Pullinger & Wagner, 2010**, p. 3). In addition, **Cox and Pinfield (2013)** emphasize that scientific libraries often do not possess specialised units experienced in both IT-skills and knowledge in domain specific research data — a factor that hinders libraries' engagement in research data management. Establishing these departments takes time and costs money — often not attractive to scientific libraries during times of budget cuts.

Based on Lyon's suggestions to assign roles between libraries and RDCs (**Lyon, 2007**; adapted by **Vlaeminck, 2013**), we suggest the following tasks for the implementation of our project's pilot application of a data archive for economics journals, in which we strive to realise a workflow based on this division of complementary know-how.

In this distribution of tasks, ZBW — the Leibniz Information Centre for Economics — adopts the role of hosting and maintaining the metadata catalogue. Libraries then provide the technical implementation of APIs to other (library or research data) catalogues with the purpose of enriching and disseminating metadata. However one of the German RDCs, the research data centre of the Socio-Economic Panel (RDC

SOEP), should take over the tasks of hosting, storing and preserving the data that has previously been submitted by editorial offices using the project's application.

By developing, implementing and operating a publication-related data archive for economics journals, both libraries and RDCs would help to ensure the validity of published economic research and to facilitate replications of these scientific outputs.

ACKNOWLEDGMENTS

The findings presented in this article have been achieved in the course of the EDaWaX (European Data Watch Extended, www.edawax.de) research project. EDaWaX is funded by the German Research Foundation (www.dfg.de). The institutions listed hereafter are involved in the project: The German Data Forum (RatSWD), the institute Inno-tec of the LMU Munich in cooperation with the Max Planck Institute for Intellectual Property and Competition Law (IMPRS-CI) as well as the German National Library of Economics / Leibniz Information-Centre for Economics (ZBW).

In addition to the authors the following persons are involved in the EDaWaX project: Professor Klaus Tochtermann (ZBW), Professor Joachim Wagner (Leuphana University, Lueneburg), Professor Dietmar Harhoff (IMPRS-CI and MCIER), Doctor Brigitte Preissl (ZBW), Patrick Andreoli-Versbach (IMPRS-CI), Doctor Frank Mueller-Langer (IMPRS-CI and MCIER), Olaf Siegert (ZBW), Ralf Toepfer (ZBW) and Doctor Hendrik Bunke (ZBW).

In particular the authors thank Adam Lederer (DIW Berlin) for valuable comments and suggestions for improving our English.

NOTES

[1]The ALLBUS (German General Social Survey) collects up-to-date data on attitudes, behaviour, and social structure in Germany. Since 1980 representative cross-sections of the population are surveyed by GESIS every two years using both constant and variable questions. Cf. http://www.gesis.org/en/allbus/allbus-home/.

[2]The German Socio-Economic Panel Study (SOEP) is a wide-ranging representative longitudinal study of private households in Germany, in which currently about 30,000 adults living in about 15,000 households are interviewed by TNS Infratest Sozialforschung on behalf of German Institute for Economic Research (DIW) Berlin each year. Cf. http://www.soep.de.

[3]Indeed, various reports and legal opinions on research data handling have been published in recent years, but it remains questionable if the uncertainty on the part of researchers has thereby been reduced [**De Cock Bruning, van Dither, Jeppersen de Boer, & Ringnalda, 2011**; **Guibault, & Wiebe, 2013**; **Häder, 2009**;**Hillegeist, 2012** (especially chapter A)].

[4]The website of dataverse can be found at: www.thedata.org.

[5]The website of the German Data Forum can be found at:http://www.ratswd.de/eng/index.html.

[6]The website of CESSDA can be found at: http://www.cessda.org.

[7]In the meantime, ICPSR's publication related archive changed its name to "replication datasets."

[8]A list of all journals and articles in which data stored at the ICPSR-PRA is included are available athttp://www.icpsr.umich.edu/icpsrweb/ICPSR/biblio/journals?collection=DATA.

[9]The Website of NARCIS, the National Academic Research and Collaborations Information System of the Netherlands can be found at:http://www.narcis.nl/about/Language/en.

[10]More Information on NARCIS can be obtained athttp://www.narcis.nl/content/pdf/narcisflyer_en.pdf. According to DANS NARCIS currently contains more than 1,800 enhanced publications and 25,000 datasets.

[11]The Website of DANS EASY can be found on: https://easy.dans.knaw.nl/ui/home.

[12]Useful information for instance was provided by **Polhout (2012)**.

[13]Respondents had the opportunity to inspect several examples of publication-related data submissions and their elements within the online questionnaire.

[14]The required elements depend on the type of research. The data availability policy of the American Economic Review (AER) – available athttp://www.aeaweb.org/aer/data.php – exemplifies such requirements.

[15]Erratum: In our survey we considered XML as a metadata schema. XML is not a metadata schema but a markup language derived from SGML. The purpose of XML is to define a set of rules for exchanging a wide variety of data. Therefore we no longer include these answers given by our respondents in our reporting.

[16]Website of nesstar, www.nesstar.com.

[17]Cf.http://www.reedelsevier.com/mediacentre/pressreleases/2010/Pages/elsevier-and-pangaea-take-next-step.aspx.

[18]Cf. http://datadryad.org/pages/integratedJournals.

REFERENCES

1. Anderson, R., Greene, W.H., McCullough, B.D., & Vinod, H.D. (2008). The role of data/code archives in the future of economic research. Journal of Economic Methodology, 15(1), 99–119

2. Baykoucheva, S. (2011). What do libraries have to do with e-Science? An interview with James L. Mullins, Dean of Purdue University Libraries. Chemical Information Bulletin, 63(1), 45–49

3. Christensen-Dalsgaard, B. (2012). Ten recommendations for libraries to get started with research data management. Final report of the LIBER working group on E-Science/Research Data Management. Retrieved January 30, 2014, from http://www.libereurope.eu/sites/default/files/The%20research%20data%20group%202012%20v7%20final.pdf

4. Costas, R., Meijer, I., Zahedi, Z., & Wouters, P. (2013). The value of research data – Metrics for datasets from a cultural and technical point of view. A Knowledge Exchange Report. Retrieved January 30, 2014 from www.knowledge-exchange.info/datametrics

5. Cox, A.M., & Pinfield, S. (2013). Research data management and libraries: Current activities and future priorities. Journal of Librarianship and Information Science. Published 'online before print', June 28, 2013.

6. De Cock Bruning, M., van Dither, B., Jeppersen de Boer, C.G., & Ringnalda, A. (2011). The legal status of research data in the Knowledge Exchange partner countries.

7. De Waard, A. (2012). Linking data to publications: Towards the execution of papers. In: Uhlir, P. (Ed.), For attribution – Developing data attribution and citation practices and standards: Summary of an International Workshop (pp. 157–159).

8. Fecher, B. (2014). Data sharing angst – An insight to an ongoing research on data sharing in academia. Alexander von Humboldt Institute for Internet and Society. Retrieved January 30, 2014, from http://www.hiig.de/en/data-sharing-angst-an-insight-to-an-ongoing-research-on-data-sharing-in-academia/

9. German Council of Science and Humanities (Wissenschaftsrat) (2012). Empfehlungen zur Weiterentwicklung der wissenschaftlichen Informationsinfrastrukturen in Deutschland bis 2020 (No. Drs. 2359-12).

10. Guibault, L., & Wiebe, A. (2013). Safe to be open. Study on the protection of research data and recommendations for access and usage. Universitätsverlag Göttingen.

11. Häder, M. (2009). Der Datenschutz in den Sozialwissenschaften. Anmerkungen zur Praxis sozialwissenschaftlicher Erhebungen und Datenverarbeitung in Deutschland. RatSWD Working Paper Series, 90, Berlin.

12. Harvey, D.R. (2012). Preserving digital materials (2nd ed.). Berlin: De Gruyter

13. Hillegeist, T. (2012). Rechtliche Probleme der elektronischen Langzeitarchivierung wissenschaftlicher Primärdaten, Göttinger Schriften zur Internetforschung (8).

14. King, G. (1995). Replication, replication. PS: Political Science and Politics, 28, 443–499.

15. Kroes, N. (2012). Commission recommendation of 17.7.2012 on access to and preservation of scientific information [No. SWD(2012) 221 final].

16. Lyon, L. (2007): Dealing with data – Roles, rights, responsibilities and relationships. Consultancy report.

17. Lyon, L. (2012). The informatics transform: Re-engineering libraries for the data decade. International Journal of Digital Curation, 7(1).

18. McCullough, B.D. (2009). Open Access economics journals and the market for reproducible economic research. Economic Analysis and Policy, 39(1), 117–126

19. Piwowar, H.A., Day, R.S., & Fridsma, D.B. (2007). Sharing detailed research data is associated with increased citation rate. PLoS ONE 2(3): e308.

20. Polhout, M. (2012). Deposit instructions for social and behavioural sciences.

21. Pullinger, J., & Wagner, G. (2010). On the respective roles of national libraries, national archives and research data centers in the preservation of and access to research data.

22. Reilly, S., Schallier, W., Schrimpf, S., Smit, E., & Wilkinson, M. (2011). Report on integration of data and publications.

23. Research Information Network (2011). Data centres: their use, value and impact. A Research Information Network report.

24. STM & ALPSP (2006). Databases, data sets, and data accessibility – views and practices of scholarly publishers. A statement by the International Association of Scientific, Technical and Medical Publishers (STM) and the Association of Learned and Professional Society Publishers (ALPSP).

25. Vlaeminck, S. (2013). Data management in scholarly journals and possible roles for libraries – Some insights from EDaWaX. LIBER Quarterly, 23(1), 48–79.

6

CHAPTER

Library Roles in University Research Assessment

John MacColl

European Director, RLG Partnership, OCLC Research, University of St Andrews, Scotland, UK

ABSTRACT

This article summarises the findings of two OCLC research reports which recently documented how university research is assessed in five countries and the role research libraries play in the various schemes. Libraries' administrative role in supplying bibliometrics is the most obvious. However, the author advocates a much more strategic role for libraries: to focus on the scholarly activity all around the library, to curate, advise on and preserve the manifold outputs of research activity.

Keywords: university research assessment, research libraries

Lorcan Dempsey wrote a blog entry a few years ago in which he stated that 'the network reconfigures the library systems environment' (July 2007).[1] Within OCLC Research we divide our research activity up into programmes, and the one that has recently been concerned with library roles in research assessment is 'Research Information Management'. I like to use a variation on Lorcan's comment here to capture what I consider to be the essence of the Research Information Management programme — 'the research environment reconfigures the library'.

RIM OVERLAPPING ENVIRONMENTS

In setting up this programme we spent some time thinking about researcher environments. Our starting point was that, in order to achieve the understanding needed to provide support for research, we had to see the world with the researcher at the centre. Accordingly, we came up with this model.

What the researcher does that concerns us as librarians is to produce outputs. Researchers inhabit a number of environments, which overlap with each other in different ways according to different national cultures, disciplinary cultures and funding regimes. The environment that matters most to academics, and is closest to their interests, is that of their domain. Here the scholars can do the things that come naturally to them. This environment is the one that provides for academic freedom. In a perfect world, scholars would inhabit this domain only, and be sufficiently remunerated that all they cared about was being a scholar — doing research, breaking new ground, gaining reputation and credit in their field, and working their way up the academic career ladder through professorship and readership emeritus to the Nobel prize. But of course, life is not that simple. Researchers depend on funding in order to support their work, and so the environment created by research funders impinges upon them. Funding bodies, whether governmental or private, whether charities or commercial companies, will each have their own mission into which the researcher's grant proposals must fit. They will often want to maintain their own records of the publications that emerge from the projects they fund, and the researcher may have to furnish details of these, or even full-text copies of their final published out-puts, into repositories or databases — perhaps with the aid of their institutional libraries. They may even have Open Access mandates now, as — for example — do the NIH in the US, and the Wellcome Trust in the UK.

Then of course the institution has a major impact on researchers. Like it or not, they depend on their institution for their monthly salary — plus a number of intellectual benefits. The institution will also have its own mission, and — increasingly — will wish to maintain its own record of publications. It may even now have an Open Access mandate.

Finally, we come to the assessment environment, which was what we studied in a project undertaken within our Research Information Management programme last year. This environment really only applies in countries in which a high proportion of the total research budget — or the HE budget from whichresearchers at least are funded — comes from the public purse. For our study, therefore, we chose five countries in which that was the case: the Netherlands, Denmark, the UK, Ireland and Australia. With a considerable amount of public money being spent on university research, there is a growing sense of the need to account for the expenditure, and to demonstrate the value of that research to the taxpayer. The assessment regime is therefore characterised by an emphasis upon national economic performance. Governments want evidence that their investment is helping their country, or region, to compete economically with their competitors in global markets. Related to that is the national research profile of a territory. What are its strengths, and where does it sit in the pecking order?

Thinking about how these various environments deal with researcher outputs, we can see that in the environment of the domain, social networks increasingly play a role — though not in all domains of course. A network like 'Nature Network'[2], however, for bioscientists, is a place where researchers can come together virtually and seek to drive up their academic reputations through informal group discussion about their latest research and linking to copies of their papers and drafts. 'Mendeley'[3] is an interesting new cross-disciplinary web-scale social tool for researchers. 'Mendeley' acquires data about publications in a whole range of hidden ways (largely by detecting it on the hard disks of academic subscribers in the way some streaming music services acquire their members' music data). It is growing organically in a way that positions it interestingly against the top-down, mandate-led, frustratingly slow approach of the academy itself. For that reason, its coverage is currently very lop-sided, with biological sciences and computer and information science the most popular groupings. The social assessment value at that scale is quite attractive however (it claims 27.5 million articles at the time of writing). The most read articles are presented in a constantly changing display on the website. It is riddled with errors, but its approach is fresh and interesting. A service like 'Mendeley' may prove to have a gravitational force on the web that in the end universities cannot ignore — and to provide a source upon which assessment tools and services are based.

In the environments of the domain, the institution and the funder, of course, we find repositories. A repository like UK PubMedCentral[4] is a domain repository, though with a funder mandate behind it in the case of the Wellcome Trust. Not all subject, or domain repositories, also serve funder requirements. We are by now familiar with institutional repositories, since they tend to be run by libraries. In the general repository configuration overall we see a lot of replication going on within a system which might be said to lack coherence. The repositories concerned each have different degrees of gravitational pull, as you would expect. Domain repositories tend to be the most attractive of all — but that does not necessarily mean they are the best populated in all disciplines. The institutional repository is unlikely to have much affinity appeal to researchers, but if heavily policed, and if it comes with a lot of mediated labour, it may still attract more deposits than the subject repository. Looking at this generally incoherent architecture, it appears to be in need of rationalisation to promote a more efficient system overall.

In the intersection of domain, institution and assessment, various factors influence and constrain researcher behaviour. Assessment requirements are usually applied by institutions, which are exercised by the degree to which any assessment regime in force shares out the funding pot which is based upon it, if there is one. The UK has the most extreme example of this, with £1.6b per year being disbursed at present to the sector as a whole, on the basis of a formula derived from how well individual institutions performed in the last Research Assessment Exercise (or RAE). Assessment results can influence league table positions, which also matter increasingly to universities in a tough international marketplace where researchers vie for grants, and institutions vie for students and the best academics.

In countries that operate tenure, it too plays a part in constraining academic behaviour. It is applied by the insitution, but always — of course — with an eye to the metrics that will be found to judge universities against each other — even without highly visible research assessment regime results. And internal assessment goes on all the time, whether as part of tenure systems, or academic staff appraisal, or as 'practice runs' for national assessment exercises.

Research Assessment and Research Libraries

Last year we commissioned Key Perspectives to undertake our study. Their report, A Comparative Review of Research Assessment Regimes in Five Countries and the Role of Libraries in the Research Assessment Process[5] appeared in December 2009. It sought to investigate the characteristics of research assessment regimes in the five different countries and to gather key stakeholders' views about the advantages and disadvantages of research assessment. It provided some analysis of the effect of research assessment procedures on the values of the academy. We were concerned not just to document the practices and procedures. There is an underlying issue here about the nature of scholarship which is surfaced by assessment in its various forms, on which we wanted to provide a library perspective. The report sought also to reveal the characteristics of research library involvement in research assessment support, and to discover the extent to which research assessment forms part of institutions' strategic planning processes. Finally, it tried to draw out points of good or best practice for libraries in support of national or institutional research assessment.

In January, we followed it up with a companion report, Research assessment and the role of the library[6], which provided a summary of the key findings of the study, with some context for the recent increase in library involvement in research assessment, and recommendations for research libraries.

University League Tables

Research assessment serves several purposes. Reputation is critical to universities as well as to the academics who work in them, as the marketplace for higher education intensifies on a global scale. The ratings obtained for research feed in to university league tables, of which there are an increasing number. This happens both directly, in some cases, where scores from a public assessment scheme like that in the UK are factored in to formulae used to generate tables, and indirectly, by affecting the perceptions of scholars whose informed views are garnered via peer review.

Perhaps the best-known league table is the listing produced annually by the UK's Times Higher Education magazine. International in scope, it is currently in the throes of a major overhaul to increase its validity. The magazine has realised that there is real competition emerging now between these various league tables. It is often criticised for being too 'UK-centric', and it will be interesting to see if that is corrected after its refreshment is complete. Its major rival, the Academic Ranking of World Universities, is produced by Shanghai Jiao Tong University. This table is better regarded in several countries, notably the US, which consider it less biased than that of the Times Higher Education. The Times Higher Education's latest table puts four UK universities (and six US universities) in the world's top 10, for example. The Academic Ranking of World Universities, on the other hand, puts only two UK universities in the top 10 (Cambridge and Oxford), and the other eight are all American.

Then there are a number of national rankings, generally produced by newspapers. The Complete University Guide is compiled in partnership with the UK broadsheet The Independent. RAE research assessment scores feed directly into its ranking order. These RAE scores, which used to come out in a single table, appeared at the last incarnation (in 2008) with a disciplinary breakdown. Nevertheless, some publications (such as Research Fortnight) managed to synthesize the results into derived league tables fairly quickly after they were published. Recently the Times Higher Education, seeking continually to enforce its position as the most authoritative league table compiler, began to produce a 'table of tables', based on three of the major UK broadsheet newspapers.

The US has also been producing league tables for many years, and the US News and World Report tables are studied anxiously by university presidents each year. The US HE system is more diversified than in most of Europe, however, so they break their rankings down into various categories — graduate schools, 'national universities', liberal arts colleges, etc.

The production of these league tables at regular intervals and the often impassioned responses to them by university managers, illustrate both the highly competitive nature of global higher education and the limitations that exist upon measures of complex organisations such as universities. University managers are caught up in agonies of anxiety over how their institutions will fare when the tables are published, and whether their position will suffer either through errors in the data used or through a misunderstanding by the compilers or by any of the data sources used in the compilation, over many of which they have no control. Careers can be damaged or broken by failures to meet university aspirations to improve their ranking positions, or by sudden reversals in position. Academic concerns for scientific rigour are affronted by what can seem to be an annual circus in which they have to compete — and yet the very fact of competition is what makes these tables matter, however much academics may dismiss them, and however many indignant letters are published in the press after each table's publication, complaining about misrepresentation and containing pleas in mitigation. This is why the Times Higher Education, having published its league tables for many years with claims to authoritativeness, has recently decided it requires to revise its compilation procedures quite drastically in order to make its table more reliable, and is doing so with a great trumpeting of virtue — an irony which will not have gone unnoticed by university managers who have suffered perceived injustices at its hands over the years.

Assessment Criteria Relativities

We found that research assessment in all of the countries we studied is composed of a mix of quantitative and qualitative criteria, and the mix tends to differ according to whether countries are prepared to spend a lot of money on the assessment exercise or not. On the qualitative side are peer review — generally considered the most important element in research assessment (look, for example, at the way the Times Higher Education is seeking to make it more rigorous in its ranking compilation at present), and 'measures of esteem' (as they are known in the RAE). These include academic prizes won, prestigious editorial positions, etc. On the quantitative side are, of course, citations and their conversion into impact factors. There are also output venue rankings — i.e. league tables of journals, and sometimes monograph and textbook publishers, where publication in these venues attracts a differentiated score which can be fed in to a formula. And finally, volume indicators —

e.g. numbers of PhD or postgraduate students, amounts of external research funding won, etc.

What we found was that the more a country wishes to save on the overhead costs of an assessment regime, the more it tends towards the quantitative measures, including bibliometric measures. Thus, when the UK set out to reform its RAE system a few years ago, it wanted to drive down the cost of the exercise by introducing a large element of bibliometrics. This was opposed and to some extend discredited, however, during the pilot testing last year, and in the end we seem to be back with virtually the same system we started with, with a strong emphasis on peer review. The amount of funding at stake for institutions as a result of these exercises is a major factor. The UK system provides much more funding to the winners of the RAE than any of the other regimes we looked at, and so it would seem likely that a high pay-out system would seem worth the high cost of credibility in the eyes of those who stand to gain or lose from it. Thinking about our environments model (Figure 1), what this tells us is that the domain values are asserting themselves by choosing peer review over other criteria. In truth, though there are always complaints, it does seem to be generally accepted by the academy, whereas trying to change the balance towards a more quantitative basis was not.

In our study we found that two of the countries surveyed took a fairly strong line on bibliometric indicators. Both Australia and Denmark are seeking to apply systems that are relatively low-cost (compared to the UK approach) and based largely on bibliometric indicators. Accordingly, the Danes have already developed — and the Australians are developing — a ranked list of publication venues (in Denmark known as the Bibliometric Research Indicator). The Australian system

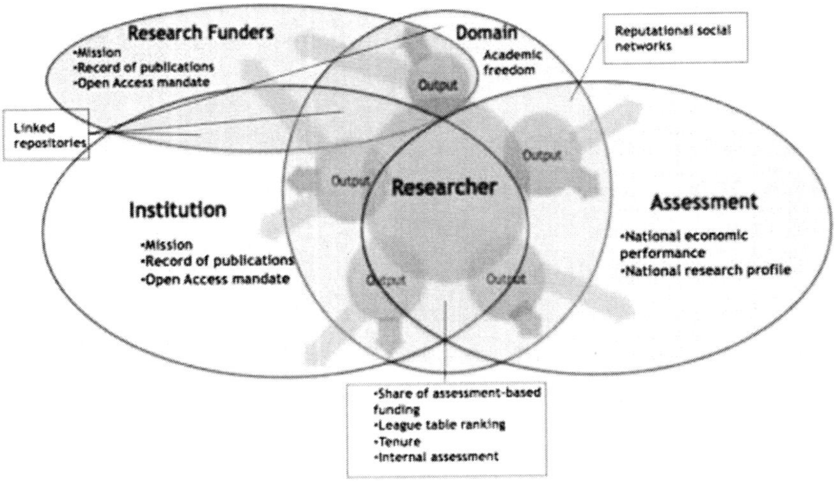

Figure 1. Research environments model.

is limited to journals, but the Danish system includes book publishers as well. Outputs are awarded points on the basis of the 'strength' of the venue in which their work appears, and co-authorship leads to elaborate fractionalising of points according to a precise formula. One can see why this approach could be contentious if there were large sums of money riding on it. However, there are indications that the Australian system will eventually determine the share-out of block grant funding of research. It will be fascinating to see if the bibliometric component can remain strong in that event.

Interviews with academics in several recent studies suggest that they are frequently skeptical about the application of numerical values ascribed to publication venues by external bodies — whether based on impact factors or on peer review. On the basis of a survey of researchers' dissemination practices undertaken last year, the Research Information Network said this about UK researchers:

'We have already noted ... the perception widespread among researchers that the RAE and the related policies of their institutions put pressure on

them to publish in journals with a high impact factor rather than in other journals that would be more effective in reaching their target audience, or to use other channels altogether. Some researchers seek actively to ignore such pressures, but many view the RAE as a game they are forced to play'.[7]

In the UK, the Times Higher Education — which recently announced a partnership with Thomson Reuters over the bibliometrics it uses in its league tables — reported in June 2010 that UK researchers were now being cited almost as often as those in the US. This finding immediately drew some criticism from academics:

'Graham Woan, reader in astrophysics at the University of Glasgow, warned that citation statistics could be skewed: "The metric used to project the raw data on to a ranking does much to define the rank order, and it is maybe only loosely coupled to the true scientific vigour and quality of a country."'[8]

This comment captures very concisely the dilemma at the heart of assessment. How do we support and sustain the 'true scientific vigour and quality of a country' while yet allowing it to be measured by those who pay for it?

Scientific vigour and quality are safeguarded within their domains, and domain values are clearly suspicious of impact factors, which are manipulable and too narrowly based on citations. Citations measure popularity, but do not consider, for example, marginal areas of a discipline which may still be important, or worthy of nurturing, even if there is a relatively small number of researchers active there.

Despite this, metrics do matter. They are useful to institutions to provide snapshot research profiles of individuals, groups and whole institutions. Because they are used in the compilation of league tables and play a part — to greater and lesser degrees as we have seen — in national systems of research assessment, it makes sense for institutions to have bibliometric expertise on site, and the library is the obvious location. A Research Information Network study undertaken jointly with OCLC Research recently found that the University of Leicester has recruited a professional bibliometrician, who will be based in the library. They were adamant that they wanted a research statistician as opposed to a librarian-turned-bibliometrician.[9]

We undertook this study during 2009, when the successor scheme to the RAE, the Research Excellence Framework (REF) pilot was running, and it was not at all clear that the outcome would be a turning aside from the government's much-publicised claim to move towards an emphasis on bibliometrics in the REF, as in fact then occurred. Had it not, we might have seen many more institutions make the same decision as Leicester. On the other hand, it does seem sensible to have such expertise in-house, ideally in the library. Though the relativities may in the end suggest that for the UK this is not such a critical skillset to have, in reality bibliometrics will continue to play a large part in assessment, and in league tables. Perhaps one of the main reasons to have a bibliometrician is not just to be able to model possible bibliometrics-based funding outcomes, but also to be able to present the institution with the drawbacks and limitations, as well as the benefits, of bibliometrics in the overall assessment mix.

ASSESSMENT AND THE RESEARCH 'ARMS RACE'

The UK RAE system has worked over the last couple of decades to concentrate research funding in the hands of a top group of universities. This can be a contentious use of public money, because it creates a 'two-tier' élite system funded at public expense. However, if that leads to a more competitive research sector for the country as a whole, the consensus seems to be that it is worth the democratic discomfort. In our report we quote Lambert and Butler in a report from 2006 that points to this deliberate UK strategy:

'One obvious explanation for the relatively high showing of UK universities in the league tables is that its research funding is much more heavily concentrated on the top institutions than elsewhere in Europe. Well over three-fifths of public and business investment in university research in England is directed to the top 15 universities'.[10]

The UK has in fact engineered its university research capacity, with the assistance of the RAE, to allow the country to compete better with the US, which has been the dominating force in world research and its accompanying economic benefits for some considerable time. If we use the analogy of an 'arms race', the US achieves its dominance not through wielding more weapons, but through wielding fewer, much more powerful ones:

'In Europe, on paper there are 2,000 universities focusing in theory on research and competing for people and funding. By comparison, in the US there are only 215 universities offering postgraduate programs, and less than a hundred among them are officially considered "research intensive universities".'[11]

The US system is of course very different. Some universities are considered 'public' (though the proportion of public funding they receive is generally much lower than would be true in Europe). Many are private. Because of this mixed economy, there is no national system of research assessment. Nevertheless, there is one major productivity driver in the US university system and that is tenure — the system that permits academic staff to reach a point in their careers where they are safe to pursue their work knowing they have a 'job for life' (from the point of achieving tenure scholars in principle need only be concerned with the demands and exigencies of their domain environment). The requirements for tenure vary considerably institution by institution. Many academics never achieve it. Others are fairly mature before they do. But inasmuch as tenure is a form of researcher assessment, it provides an interesting parallel to our study of research assessment regimes. Might libraries in US universities play a role in supporting the tenure and promotion system that is akin to the one we consider them playing in our publicly funded regimes in support of research assessment?

SKEW DRIVERS

We have discussed the Danish, Australian and UK systems, and can see that the assessment regime has a strong impact on behaviours in these countries. In a sense what is happening is that the freedom of researchers to do the work they want to do is being compromised and their outputs are necessarily being skewed towards the assessment environment. In the UK that means favouring journal publication over other forms, since journal articles are more quickly produced than monographs and so can be counted more readily. In the case of Denmark and Australia, the assessment regime's requirement to use nationally produced tables of output venue rankings inevitably skews behaviour in ways one can readily see are likely to be contentious. There is a danger that the tables ossify what should be a dynamic disciplinary literature. They also skew publishing behaviour and are biased against new journals or publishers entering the field.

When we looked at the Netherlands, however, we found a different picture, in which assessment was performed largely on an institutional basis. The absence, to date, of a government agenda to concentrate research for reasons of economic competitiveness — in the Netherlands and indeed in much of Europe — has allowed a traditional culture of university self-regulation to continue. There are signs, however, that that may be changing. Ireland as yet has no system of assessment country-wide, and so it too presents the institutional environment as the one with the most influence.

Returning to our research environments model in the context of the US tenure system, it seems clear that its key environments are those of the domain and the institution. Tenure is achieved through a form of internal peer review, which is highly concentrated in that it applies researcher by researcher. While reputation in the wider field of one's discipline is of course critical, achieving tenure in one's own institution is an even greater pressure — and the one feeds into the other. This is clearly brought out in the recent Faculty survey by Ithaka S+R in the US:

'For most faculty members, our data seem to be consistent with other research indicating that faculty interest in revamping the scholarly publishing system is secondary to concern about career advancement, and that activities that will not be positively recognized in tenure and promotion processes are generally not a priority' (Ithaka S+R Faculty survey 2009).

What is said here about the US in the context of tenure could equally be said about other countries (such as the Netherlands) with tenure systems, and indeed about the countries in our study which have abolished tenure — such as the UK. Tenure may not exist as a formal system, but the effects it creates exist nonetheless in the career-building culture of the academy. Reform of scholarly publishing, open access, the population of repositories and other such high library priorities are very much second order activities for academics.

Tenure exerts a strong pressure on the scholar as individual. The recent highly publicised murder of three professors by a lecturer denied tenure at the University of Alabama brought this comment from a senior academic:

'As much as it may be uncomfortable for us to admit and discuss it, we in academe need to confront the psychological and physiological effects of our culture on our rising scholars. We have done a ... poor job of providing the support and mentoring appropriate for such major, stressful, career make-or-break situations as dissertation defenses and tenure votes ... It is time to end any tolerance for the notion that "we eat our young" and that such intellectual brutality is somehow an indicator of rigor.'[12]

Thus, while it may embody domain values, the institutional expression of those in highly competitive universities can be oppressive. Some 45% of sociologists surveyed by Ithaka think that tenure creates deleterious skew in their dissemination options.

The skew in the UK is caused by the RAE:

'Many researchers believe the current environment puts pressure on them to publish too much, too soon, and in in-appropriate formats ... researchers are increasingly aware that publications serve not only as means of communication. They can be monitored or measured as indicators of quality or impact ... the perception that their work is being monitored and assessed, by the RAE in particular, has a major influence on how researchers communicate'.[7]

The suggestion that the RAE encourages 'game-playing' appears in several places throughout the Research Information Network report. To emphasise the parallel effect in the US created by tenure, the following quote comes from a researcher surveyed as part of a recent comprehensive study by the Center for Studies in Higher Education at the University of California Berkeley:

'The increasing need to demonstrate progress on a second project ... may encourage a certain amount of "games-playing" to the detriment of good scholarship ... So that, at a place like this (not Harvard, Yale, or Princeton), they're not asking for the second book yet for tenure in history, but they are asking more than they used to for signs of how far along are you on the second book. And frankly, I think it encourages people to fake it'.[13]

Though both systems exhibit skew, and cause unhappiness to researchers as a consequence, there are some considerable differences in the effects. One of the most significant relates to the scholarly place of the monograph. In the US, as we have seen, tenure can be held back in some institutions until researchers have completed a second book. In the UK, by contrast, humanities scholars complain that the system does not value the monograph as they do:

'In this environment, articles in scholarly journals are increasing their dominance over all other forms of publication and dissemination ... journals and the articles they contain are also the form of publication most easily measured, ranked and assessed, and thus most used in the measurement of research performance ... while some researchers in the humanities and social sciences complain of inappropriate pressure to publish articles rather than books, a majority feel that monographs remain the single most important mode of dissemination, one around which they build their careers.'[7]

In the UK, the charge against the RAE is that it encourages game-playing by whole departments, or whole institutions — particularly in the area of the star researcher 'transfer market', but also in areas such as choice of discipline subcategory to enter. In the 2008 RAE, the Unit of Assessment Library & Information Management, for example, caused some surprise in the library community by including King's College London (which does not have a library school, but submitted its Digital Humanities unit in this category) and excluding Strathclyde University, whose School of Information Science was merged into a new Department of Computer and Information Sciences following the previous RAE in 2001.

WHAT DOES IT MEAN FOR LIBRARIES?

At the recent RLG Partnership Annual Symposium in Chicago, Adrian Johns — who gave a keynote address entitled 'As our readers go digital' — said in conclusion that the role of librarians in relation to academics needs to be remade in a more permeable way. He spoke of the role of 'intelligencers' in the days of the 'Republic of Letters' — people who acted as conduits for scholarly knowledge among scientists and philosophers, who advised scholars of the latest publications and findings in their field. This is a theme we are pursuing elsewhere in our Research Information Management programme, looking specifically at the role of subject librarians — or what Jim Neal of Columbia University in a panel at the ACRL conference in 2009 called 'Subject Librarian 2.0'.[14] In thinking about assessment, it is clear that libraries have a number of roles to play in the processes that exist in the national regimes for assessment. Most obviously, insofar as these systems (and tenure and promotion systems in the US) require correct bibliographic metadata to operate, the library is the obvious place to get it.

But there is a more strategic role that we librarians can and should play here, and it is one which I would suggest has been weakened over the past two or three decades of technological progress as libraries have embraced technologies that have concentrated on data processing and so allowed them to play important roles in university administration. Where overall funding has been more constrained, their movement in this direction has been at the expense of their scholarly curation activities and they now need to reassert their role in respect of scholarly knowledge. The library should be knowledgeable about knowledge, and should be the main authority on the campus about the ways knowledge is generated and transmitted through all of the disciplines it contains. The library is the only neutral scholarly actor on the campus. Certainly it assists with administration too, and much of what we discovered in this project is that the administrative role is the main one performed by the library in relation to assessment. But there are a number of expertises that are not obviously discoverable on most university campuses today, and yet which belong in this place of 'neutral scholarship'. They would include expertise in bibliometrics, in copyright and licensing (and thus in open access), in publishing, and in the tools of scholarly dissemination (the blogs that are most useful in particular fields, etc). And these expertises, ideally located in subject liaison librarians, collectively should be represented by the University Librarian and senior colleagues, to whom the deans and principals of our institutions of Higher Education should turn regularly for advice, and from whom they should hear regularly about the changing contours of scholarly outputs in the disciplines of their institutions.

There surely is no longer any doubt that institutional libraries must set out on a radically transforming path, scaling back on institutional-level cataloguing and acquisition, and on institutional library systems management. There are signs in some institutions that this is already happening. These tasks will be done increasingly by machinery and third parties. The institutional research library needs to focus on the unique materials it holds within its collections, and put them in the flow. And it needs to focus on the scholarly activity that goes on all around it on campus — more quickly, more vigorously and in greater volume than ever — and largely without reference to the library. Without the assistance of the library to curate, advise on and preserve the manifold outputs of this activity, while individual scholars may still manage to thrive and build their reputations, they will do so within an impoverished infrastructure for scholarship, using a compromised archive, and their legacy to future scholars will be insecure. Our greatest challenge now is to understand this, and to shape our professional structures to deal with it.

June 2010

NOTES

1. Lorcan Dempsey's weblog [http://orweblog.oclc.org/archives/001379.html].

2. [http://network.nature.com/].

3. [www.mendeley.com].

4. [http://ukpmc.ac.uk/].

5. Key Perspectives Ltd (2009): A Comparative Review of Research Assessment Regimes in Five Countries and the Role of Libraries in the Research Assessment Process. Report commissioned by OCLC Research. Published online at [http://www.oclc.org/research/ publications/library/ 2009/2009-09.pdf].

6. MacColl, John (2010): Research Assessment and the Role of the Library. Report produced by OCLC Research. Published online at [http://www.oclc.org/research /publications/library/2010/2010-01.pdf].

7. Research Information Network (2009): Communicating knowledge: how and why UK researchers publish and disseminate their findings [www.rin.ac.uk/ communicating-knowledge].

8. 'Britain closes science-citations gap on US', Times Higher Education Education, 10 June 2010.

9. Research Information Network (2010). Research support services in UK universities [in press].

10. Lambert, Richard and Nick Butler (2006): 'The Future of European Universities: Renaissance or Decay?' Centre for European Reform.

11. Sironi, Andrea (2007): 'European Universities Should Try to Conquer the World' Università Bocconi web site. Available online at: [http://www.virtualbocconi.com/Articoli/European_Universities_Should_Try_to_Conquer_the_World/default.aspx].

12. The Chronicle of Higher Education, February 21, 2010.

13. Center for Studies in Higher Education, UC Berkeley (2010): Assessing the Future Landscape of Scholarly Communication: An Exploration of Faculty Values and Needs in Seven Disciplines [http://escholarship.org/uc/item/15x7385g].

14. See blog post at [http://www.learningtimes.net/acrlconference/2009/subject-liaison-20/].

7

CHAPTER

Evaluation as a Tool for Developing the Quality of the Academic Libraries. Case Study at a Large and Traditional Research Oriented Scandinavian University - Viewpoint of a Member of the Evaluation Panel

Gunnar Sahlin

Kungl. Bibliotheket, National Library of Sweden, P.O. Box 5039, 1002 41 Stockholm, Sweden

ABSTRACT

We were invited as an international review team to assess the development of the library system in the Helsinki University. As reviewers and experts we came to Helsinki for two days in May 2004 - Hans Geleijnse from The Netherlands, Sinikka Koskiala from Finland and myself from Sweden. The result of our evaluation is the report Follow-up Evaluation of Library and information Services in Helsinki University 2004 (Geleijnse, Koskiala & Sahlin, 2004).

THE COMMISSION

We were invited as an international review team to assess the development of the library system in the Helsinki University. As reviewers and experts we came to Helsinki for two days in May 2004 - Hans Geleijnse from The Netherlands, Sinikka Koskiala from Finland and myself from Sweden. The result of our evaluation is the report Follow-up Evaluation of Library and information Services in Helsinki University 2004 (Geleijnse, Koskiala & Sahlin, 2004).

The university requested the panel to answer some questions, and they wanted us to give some comments on the key issues for the future. The questions and key issues were the starting point for our investigation. Another base for our study was the assessment of the library activities in the university from the year 2000 (Geleijnse et al, 2000). The strategy for the university, the assessment report with different viewpoints and recommendations and other relevant information were useful for us in the evaluation. How has the university succeeded in its library development, required by the panel from year 2000?

THE EVALUATION PROCESS

The evaluation started the year before with self-evaluations of librarians and library staff, deans, directors of independent institutes, researchers, teachers, and students. The reports from these self-evaluations were provided to us. Besides the report from year 2000 and the self-evaluations we took part of other written material, from the libraries and the university, before our visit to Helsinki. The strategy for the university, action plans for libraries and the information technology, data and statistics concerning library activities, annual reports and other relevant information were delivered to us. All this material was very useful in the evaluation.

A steering-group, chaired by Vice Rector and chairman of the Library Committee, Hannele Niemi, was formed in 2003 to prepare and co-ordinate the review. They, together with the evaluation officer and the Director of Information and Library Services, gave us good support. Besides the steering group we met the Rector and deans, groups of researchers, teachers, students, and of course library staff. The National Librarian and Directors for the faculty and campus libraries gave us their descriptions. We visited branch libraries and the educations centre, Alexandria Learning Centre.

The written material, the visits, and the group-meetings - all gave us a broad and clear view of the library situation at the University of Helsinki. Following the visit our notes and impressions blended together - in close contacts through e-mail - to the evaluation report.

Even if the evaluation is two years old, it's no problem to remember the meetings, the visits and the results. First of all I would like to express that I'm satisfied with the evaluation - the process and the report. It gave me, and I hope also the other in the evaluation group, very interesting impressions, which are useful in several of following situations. When you are part of an evaluation group you give your knowledge and your impressions. At the same time you get something back, new experiences which are useful for yourself.

The second thing I will say is about decentralization. I have worked in several universities - as researcher, teacher, library director and as member of board - and every university is by nature a decentralized organization. But Helsinki University is one of the most decentralized universities I have seen. It was important to notice this fact when we studied the situation for the library organization.

The evaluation report

In the panel we were pleased with the activities following the assessment year 2000. One example is the creation of a small central unit, responsible for cooperation and electronic resources, and it's important to have this unit permanent.

Despite the development in the university libraries from year 2000, we identified during our evaluation process some important obstacles, problems or questions. I will give you some examples of these and the decentralization, funding model, and the role of the National Library are the common denominators for these obstacles, problems or questions. A special situation for Finland was the National library's close relation to the university. They still get their financial support through the university but the situation has improved. For the other problems we scrutinized many library directors who can find some relations to their own situation:

- a continuous tension concerning what should be done on a central level and what on a decentralized level in the campus libraries,
- the management of the important differences between the campus libraries,
- the need for a solid funding model for the future,
- a coherent staff strategy,
- the cost of space and the fact the available space is partly underused,
- the need to make better use of the expensive information resources,
- clear service agreements between the University of Helsinki and the National Library.

In our study of the structure and organization of the university libraries we notice the important and strong role of the campus libraries, the new role of the National Library and the importance of the central Unit of Information and Library Services and Development. Together they form what we call 'The University Library'. In our recommendations we noticed - some examples - a need

- for further cooperation between these three parts but also between the different campus libraries,
- for a sustainable staff strategy to better use of the human resources,
- further clarifications of the relations between the university and the National Library, with tasks being performed by the National Library and service level agreements were needed,
- to increase the funding for licensing and electronic information.

Some of our ideas in the evaluation group for the development in the university and in the National Library have been fulfilled. Of course it depends on great activities from those persons responsible for the development, but for us in the evaluation group it's a pleasure to have been a small part of this development in the library of Helsinki University.

REFERENCES

1. Geleijnse, Hans, Göran Gellerstam, Ian R.M. Mowat and Kyllikki Ruokonen; introduction by Aimo Virtanen: Helsinki University Libraries - Report of an Assessment panel. Helsinki : University of Helsinki, 2000. (Evaluation Projects of the University of Helsinki, 7).

2. Geleijnse, Hans, Sinikka Koskiala and Gunnar Sahlin: Follow-up Evaluation of the Library and Information Services 2004. Helsinki, 2004. (Evalution Projects of the University of Helsinki, 8).

8

CHAPTER

Information Society and Library Evaluation Transitions in Portugal: A Meta-evaluation Model and Frameworks (1970–2013)

Leonor Gaspar Pinto, Ochôa Paula

Universidade Nova de Lisboa, Portugal

ABSTRACT

The need for greater understanding of assessment practices and models highlights a deficit of an up-to-date meta-evaluation model, whilst articulating with new phases in Information Society (IS) development. This paper aims to discuss the meta-evaluation model and frameworks that were created to explain the relations between IS transitions and the development of library performance evaluation models in Portugal (1970–2013).

The research is based on a qualitative methodology supported by a combination of literature review with the construction and application of conceptual models and frameworks.

The meta-evaluation model of the impact of transitions on library performance evaluation provides an adequate representation and explanation of relationships between IS transitions and library performance evaluation models. The CLPET

(Categorizing Library Performance Evaluation Typologies) Matrix, as well as the relational framework that was developed proved to be useful analytical tools.

This paper highlights the current transition of performance models into a more holistic performance management, clarifying that diverse uses and components of performance are strongly linked to IS dynamics. It will also enable readers to discuss the impact of meta-evaluation models as a strong instrument to support the challenges of visions, strategies and best practices shifted over time.

Keywords: information society transitions, library performance evaluation, meta-models

1. INTRODUCTION

In the last 30 years, both library assessment and Information Society (IS) developments have experienced a great and long period of theory building and conceptualization. The most striking feature of performance management is its continuing expansion over the last two decades making this period its international apogee. The 1990s are considered to be the rise of performance, with measurements becoming more intensive, extensive, external (**Bouckaert & Halligan, 2006**), professionalized and institutionalized (**Demmke, 2006**). There have been many debates over the value of performance, measurement and its management, specifications of its components and their relationships and applications in different countries, characterized by a constructive period of consolidation and refinement of measurement approaches and instruments. Several performance issues have been researched – performance appraisal, policy planning, external reporting, performance based budgeting and audit – and the crucial question has been *how to develop performance measurement in a usable and functional way as a tool for policy and management* (**Van Dooren, 2005**). Another emerging approach is the discussion of theoretical insights into how information is processed and the reasons of its non-use (**Van Dooren, 2011**). Special attention has also been paid to the way in which decision makers use performance measurement information vis-à-vis their prior understanding, experience and evidences, aligning performance models and evidences of institutional uses, namely 'good practices'. Recognizing the political nature of performance management and assuming that high quality dialogue will lead to improved judgment and decision-making may strengthen the role of performance in complex and ambiguous contexts (**Van Dooren, Bouckaert, & Halligan, 2010**).

Among the Portuguese Library and Information Services (LIS) professional and academic community some performance evaluation approaches and models became almost predominant, while other international trends failed to break through. The discussion about evaluation in Portuguese libraries has evolved from pioneering experiences and isolated cases into the notion of best practices with interest to the entire library sector. Libraries are social organizations and, therefore, they can never be static entities. As the environment changes, so must libraries. This means that library evaluation models must also keep pace with the rate of change and, consequently, they have multiple interactions with policy making transitions, suffering several moments of feedback, adjustment, learning and adaptation, all interacting in a variety of ways, risks and opportunities.

According to **Van Dooren (2005)**, a performance measurement policy is the missing link in reforms, but, in general, cultural organizations have not developed advanced performance measurement systems that highlight both cultural/economic performances and their impacts. **Bonet and Donato (2011)**explain this attitude by different factors: the intrinsic difficulties in measuring a symbolic value; the frequent inability to define consistent performance measurement systems, which results from difficulties in setting the mission and the strategic goals; and the existence of governance systems that are less oriented to the stakeholders, resulting in less attention to the external communication of performances. So, these authors conclude that there are responsibilities, as well as big challenges, for cultural politicians and for cultural management researchers: "There is an ongoing tension between contrasting forces: centralization *versus* decentralization; organizations' dependence *versus*autonomy; general objectives *versus* specific objectives; extrinsic goals *versus*intrinsic goals. In that context, the key factors to successfully mastering the crisis are: a) building decision-making processes that should be fluent, non bureaucratic and participatory, and at the same time able to combine responsibility with autonomy; b) moving towards a long term strategic approach defining policy and organizational priorities" (**Bonet and Donato, 2011**, p. 9).

Understanding libraries' evaluation related fields cannot ignore the concept of Excellence (European Foundation for Quality Management – EFQM, Business Excellence Model, 2003), closely tied to the development of quality management theories and models and to performance evaluation dynamics in European organizations. The main concepts are focused on result orientation principles, leadership, process management, development and involvement of people, continuous training, partnerships and social responsibility. Organizational maturity and sustainability of results posed new issues related to new meanings and perspectives to convey to another step. Integrated, balanced, multidimensional, comprehensive and holistic models are frameworks used as tools to achieve the excellence stage (ISO 9000; Investors in people; EFQM Excellence Model; Balanced Scorecard; SERVQUAL).

Discussions about evaluation variety are growing, with a particular interest in several research themes:

1. Quality management models and their forms of development and implementation in organizations (**Cullen, 2005**; **Oakland, 2003**; **Wilson & Town, 2006**).

2. Main tendencies about economic value1 (**Aabø, 2011**; **Almauk, 2012**).

3. Leadership styles, benchmarking initiatives and good practices (**Poll, 2007**;**Staes & Thijs, 2010**).

4. Public policies about performance evaluation (**Bouckaert, 2012**).

5. Organizational learning.

6. Critical factors of digital success that promote Excellence (**Gonçalves, Moreira, Fox, & Watson, 2007**; **Huijboom & Van der Broek, 2011**; **Plum, Franklin, Kyrillidou, Roebuck, & Davis, 2010**).

7. Evidence-based management (**Rankin, 2012**; **Tanner & Deegan, 2012**; **Town, 2011**).

8. Results and impacts to consolidate new and convergent contexts of Excellence (**Poll, 2012**; **Streatfield & Markless, 2009**; **Tenopir, 2012**).

Library assessment has been mainly about quality rather than about value (**Town, 2011**) and excellence (**Ochôa & Pinto, 2012**). In many countries, after the first performance measurement revolution, a second one is needed and will happen soon. New measures linked to institutional objectives are needed and must be communicated: holistic and coherent long-term measures that assess societal benefit, educational impact and intangible assets resources, like relational capital, organizational capital and human and intellectual capital, transferable skills, collections, services, research management, research information, meta content (**Town, 2011**) and user-defined value metrics of electronic resources (**Chew, Stemper, Lilyard, & Schoenborn, 2012**). Knowledge management and Competency management are considered important elements of organisational strategy and pillars of success and they are a welcome shift in thinking about these issues.

Measures of success are faceted across several categories (**Tanner & Deegan, 2012**, p. 42):

- *Users, audience and stakeholders* – how libraries benefit them.
- *Innovation and development* – evaluating the value of new products and services in terms of how it enables users and providers to do new and innovative actions (new research and teaching methods, web 2.0 type of interactions, new modes of collection development and curation)
- *Internal processes* – evaluating how strategy provides value to the way in which the organization operates.
- *Financial benefits and outcomes* – performance evaluation.

Value maps, relating library behaviours to end user benefits, form one of the proposals to highlight value. Value mapping goes beyond performance measurement and acts as a performance management of excellence, using and supporting assessments, planning, decision-making, communication, facilitation and evaluation.

This move, from measurement towards value and excellence, is at the centre of concomitant changes for librarians. Research will be required to determine which factors are relevant to different types of libraries in making management decisions.

Influenced by eGovRTD2020 recommendations (**Codagnone & Wimmer, 2007**) to contemplate holistic research themes, actions and actors, doctoral research carried out by each of the two authors2 led to a joint meta-evaluation3 study on the impact of main IS transitions on performance evaluation models implemented in Portuguese libraries, providing evidence for the exploration of new scenarios for theoretical development. As in other countries, in Portugal research related to the impacts of libraries is uncoordinated and the nonexistence of a national coordinating body for libraries and a national research agenda to encourage consistency in evaluation methodologies characterize the major difficulties in establishing regular performance practices. The Information Society (IS) has been developed mainly since 1997 with several governmental strategies focused on qualifications, education and training, social inclusion and accessibility; simplification and better public services, digital contents and services and ITC use on a daily basis. Many strategies followed the European Union agenda, trying to accomplish benchmarking indicators and outcomes. Some transitions in European strategy underline several impacts in institution management that must be examined. This is the case of libraries that developed models in specific contexts related to specific IS factors, using the theories and instrumentalities of evaluation, resulting in a cumulative body of knowledge and skills that must be highlighted.

This project's core research question was '*How did major Information Society transitions impact on library performance models in Portugal?*' Based on a qualitative methodology supported by a combination of literature review with the construction and application of conceptual models and frameworks, the research was anchored in two core elements: Information Society and library performance models. Digging further into potential interrelationships and exploring interactions, the following steps were taken:

1. Identification of the main IS Transitions Cycles in Portugal and their relation to the library context (academic, governmental, specialized, school and public libraries);
2. Identification of main library performance models – mapping using a categorization matrix;
3. Analysis and identification of IS impacts on the adoption of models in a holistic reference framework, mapping four main categories.

Figure 1 presents the conceptual (meta-evaluation) model that was designed as an explicative hypothesis.

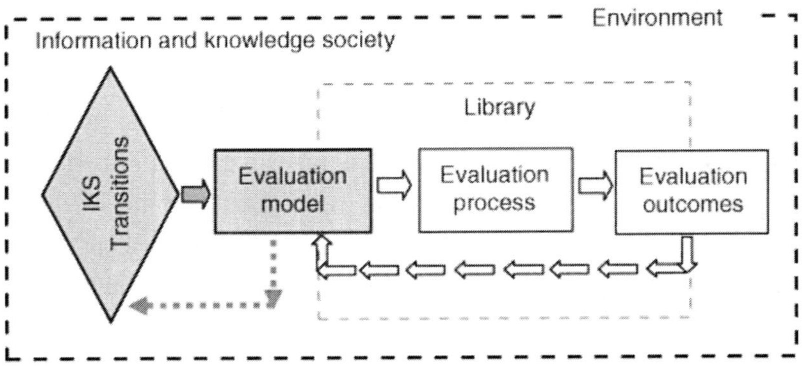

Figure 1. Meta-evaluation model of the impact of transitions on library performance evaluation models.

This research is intended to inform the debate by policy formers of four main kinds: those concerned about the overall quality of services; those concerned with institutional and sector comparisons; those within institutions concerned with interpreting their own performance data appropriately, and those who study information society developments.

The project was presented at the 3rd Qualitative and Quantitative Methods in Libraries International Conference (QQML, 2012) and has already contributed to a series of academic discussions.

2. INFORMATION SOCIETY AND LIBRARY PERFORMANCE EVALUATION MODELS IN PORTUGAL

Evaluation can be considered as an autonomous scientific discipline. In fact, it is a "transdiscipline" whose subject matter is the study and improvement of tools for other disciplines (**Scriven, 1991**, p. 365). Decades of Evaluation Research applied to Library and Information Science have resulted in a multiplicity of theoretical and empirical studies, from the pioneer work of **Bradford (1948)** to the publication of the first international standard on library performance indicators (ISO 11620, 1998) and recent developments in the emerging arena of library impact evaluation (**Imholz & Weil, 2007**; **Markless & Streatfield, 2006**;**Mays, Tenopir, & Kaufman, 2010**; **Streatfield & Markless, 2009**; among others). Within the library evaluation research *corpus*, meta-evaluation is still an underexplored area (**Calvert, 2008**; **Pinto, 2012**; **White, 2002**).

As **Calvert (2008)** points out, "Some authors have described the diversity of library evaluation methods in some detail (...) yet few have attempted to evaluate the variety of evaluation methods or to model library evaluation" (p. 2). Based on Systems Theory, this author and, before him, **Nicholson (2004)** built a cell matrix to categorize the types of library performance evaluation. Research carried out for the doctoral degree by one of the authors led to the development of their proposals and to the creation of a new and enhanced conceptual framework for categorizing types of library performance evaluation models/approachesthe **C^{LPE}T** (^{C}ategorizing ^{L}ibrary ^{P}erformance ^{E}valuati on^{T}ypologies) **Matrix**.The ategorization is framed by the Matrix's axes:

the *y*-axis, which uses three possible *Perspectives* on evaluation (*Internal, External* and *Holistic*) and the *x*-axis, which reflects the evaluation *Topic* – the *Library itself* or the *Use* made of its products and services.

This meta-evaluation approach was employed to identify main performance evaluation models and practice types used in Portuguese libraries from the 1970s to 2010, as shown in Table 1. With the exception of statistics, which in its most rudimentary form can be traced back to the late 19th century, all library performance evaluation models that were categorized in the **C**^{LPE}**T** Matrix present their creation/start date.

Pinto (2012) [adapted from **Nicholson (2004)** and **Calvert (2008)**]

Legend: **PLNN** – Public Libraries National Network **LMLN** – Lisbon Municipal Libraries Network

KRCN – Knowledge and Resources Centres Network

The longitudinal analysis of the context of library performance evaluation in Portugal highlights four main tendencies, which are reflected in the **C**^{LPE}**T** Matrix content (see Table 1):

2.1. Emergence of user primacy

In Portugal, the first reflections and studies on library use and users' profile can be traced back to the 1970s. In the next decade, albeit with less vigour than before, User Studies4 continue to dominate the small Portuguese library performance evaluation *corpus*. Although there was an active interest in studying the information use and its users in the 90s, it was only with the advent of the next century that user primacy

Table 1. Main performance evaluation models and practice types used in Portuguese libraries (1970–2010).

		TOPIC	
		Library	**Use**
PERSPECTIVE	**Internal** (Library)	Inputs Results- products Processes Goals Statistics Benchmarking Standards (1998) PLNN Model of Performance Information Management System (1993) LMLN System for collecting and reporting performance information (1991) KRCN Performance Evaluation Model (2001) ISO 9001 certification (2004)	Results – library use User Studies [focus on library use] Bibliometric studies Goals Statistics Benchmarking Standards (1998) PLNN Model of Performance Information Management System (1993) LMLN System for collecting and reporting performance information (1991) KRCN Performance Evaluation Model (2001)
PERSPECTIVE	External (Users and other stakeholders)	Needs and expectations of customers KRCN Quality Observation Model (2001) LMLN Service Quality Model (2004)	Customers' satisfaction Service Quality Impact on users Social and economic impact KRCN Quality Observation Model (2001) LMLN Service Quality Model (2004) ENTITLE-LMNL Impact Assessment Framework (2008)
PERSPECTIVE	**Holistic**	(Total) Quality Management EFQM CAF BSC Integrated Models MonitorDoc (1999) IU-ME Quality Program (1996) LMLN Performance Evaluation Integrated System (2003) SIADAP+8 Model (2004) Mix-model CAF-BSC(-AHP) (2005) Digital Library Integrated Evaluation Model (2006) School Libraries Self-assessment Model (2008)	

was enhanced, boosted by theoretical and empirical research on users' digital environment interactions, bibliometrics and impact assessment. The ENTITLE-LMNL Impact Assessment Framework Model reflects this increased attention to library impact on individuals, bringing an extended user perspective into the assessment process. This Model was tested and implemented in the Lisbon Municipal Libraries Network (LMLM) in the context of its participation in the ENTITLE (Europe's New libraries Together In Transversal Learning Environments) Project, from 2008 to 2009 (**ENTITLE, 2009**; **Pinto, 2009**).

2.2. Valorising library performance information

For quite a long time, Portuguese library performance measurement corresponded to irregular collections of input and output statistics, which were considered to be end results with a very vague connection to the management functions. This situation, which characterizes the primordial stage of library service assessment (**Shi & Levy, 2005**), started to be reversed in the early 1990s, dragged by two initiatives: the development in 1991 of the LMLN System for collecting and reporting performance information and the creation in 1993 of the Public Libraries National Network (PLNN) Model of Performance Information Management System. Both models were anchored in an internal perspective on library performance evaluation centred on both Library and Use topics. The model, as well as the development process and results of the second initiative, were presented and discussed in 1997 at the IFLA Satellite Meeting on Performance measurement and quality management in public libraries (**Pinto, 1998**) and at the 2nd Northumbria International Conference on Performance Measurement, and were the main responsible for the score 3 ("Government-backed performance measures") given to Portugal in a world-wide survey on Government interest in library performance measures (**Winkworth & Gannon-Leary, 2000**).

The first performance evaluation model specifically built for the LIS sector – the Knowledge Resources Centre Network (KRCN) Performance Evaluation Model – was conceived in 2001. Developed and implemented by OQIKS – Observatory of Quality in Information and Knowledge Services, this model was based on four types of references, one of them being the international library standards ISO 11620 and ISO 2789.

This *transition period* is characterized by the special attention given to the form and format of assessment by a typology of libraries.

2.3. Quality

The introduction of Quality Management in Portuguese libraries is directly related to the evolution of quality in Public Administration, which was strongly conditioned by the accession of Portugal to the European Union in 1986. This quality path was pioneered in 1996 by the Information Unit of the General Secretariat of the Portuguese Ministry of Education (IU-ME) through the development of three main management strategies focused on clients' needs within the context of an enduring Quality Programme (**Ochôa & Pinto, 2006**):

- 1996–2000: Fostering and adapting quality models and methods, namely EFQM – European Foundation for Quality Management's excellence model and CAF – Common Assessment Framework, the self-evaluation framework recommended for European Public Administration Services. During this period, the IU-ME team carried out two administrative modernization research projects, one of which resulted in the creation of a tailored self-assessment model for public administration information units – MonitorDoc.
- 2001–2002: Developing and implementing a quality concept linked to Knowledge Management.
- 2003–2006: Developing and implementing a quality concept linked to competency management.

If we exclude the distinctive case of the IU-ME, only at the dawn of the 21st century Portuguese LIS academics and practitioners started paying more attention to the external perspective on library performance evaluation, under the growing impetus of Quality. This trend was primarily rooted in two initiatives: the first, pursued by OQIKS between 2001 and 2003, was aimed at building KRCN's quality concept and led to the development of the KRCN Quality Observation Model; the second one was part of a wider strategy targeted at the construction of an evaluation culture in LMLN (2004–2006) and implemented through the LMLN Service Quality Model, a model adapted from Parasuraman, Zeithaml and Berry's Gap Model (**Vinagre, Pinto, & Neves, 2011**).

2.4. Integrated models

By capitalizing on previous experiences and reflections on the application of various models and management tools to the LIS field, a new trend centred on a holistic perspective on library performance evaluation came up during the first decade of the 21st century. In fact, the groundbreaking appearance of MonitorDoc (1999) was followed by the design and implementation of several integrated performance evaluation models: LMLN Performance Evaluation Integrated System (2003), SIADAP+B Model (2004), Mix-model CAF-BSC (-AHP) (2005 – **Melo & Pires, 2008**), Digital Library Integrated Evaluation Model (2006 – **Pinto, Ochôa, & Vinagre, 2009**) and School Libraries Self-assessment Model (2008 – **Portugal. Ministério da Educação e Ciência. Gabinete da Rede Bibliotecas Escolares, 2011**).

It was only after 1997 that Portugal developed the main measures towards IS, with political attention to libraries gathered from strategic initiatives in Europe, library programmes and national developments of library automation. Over time and with guidance from e-Europe Action Plans, library services were measured and integrated in sectorial actions to develop citizenship, informational and digital literacy, e-science and digital libraries. Different phases were developed in order to implement an IS agenda, with cycles of technological investment and of modernization initiatives. In periods of technological changes, libraries were a fertile area for generating new services for citizens. In periods of modernization, libraries were faced with strategic changes and some uncertainty and risk in their outcomes. Some researchers see modernization cycles as exponential transformations in evaluation practices and a predominant feature of public services crossroad with New Management, influencing the choice of models and pathways.

The focus on meta-evaluation implies that researchers not only examine library actions, but also their alignment with European orientations and the meanings that such actions have for LIS professionals in underpinning their choices, interactions and use of models. This implies that an integrated micro/meso/macro performance focus is needed due to the reported difficulties for integrating this information in a comprehensive, coherent, consistent and functional way in policy and management cycles (**Bouckaert & Halligan, 2006;Demmke, 2006; Van Dooren, 2011**).

Our research was designed to meet these challenges and led to the identification of four major transitions in IS library environmental scanning:

- *Transition I* – started with the Schwencke Resolution and its impact on national library policies;
- *Transition II* – started with the Bangemman Report and its impact on national IKS policies;
- *Transition III* – started with the Lisbon Strategy and its impact on quality strategies and on the e-Europe agenda;
- *Transition IV* – started with the European Digital Agenda, focused on new trends for services to citizens and enterprises.

3. TRANSITIONS AND IMPACTS ON LIBRARY PERFORMANCE EVALUATION MODELS IN PORTUGAL

Aiming at a better understanding of the relationship between transitions and performance evaluation models, the **C^{LPE}T** Matrix was combined with another approach – Bouckaert and Halligan's *Four models of performance* (2006) – that represents the historical development of performance and management as a basis for analysing and comparing country orientations to performance in four components: *measuring, incorporating, using* and *limitations.* According to **Bouckaert and Halligan (2006)**, contextual analysis of different performance management systems is a significant element since it enables comparisons across countries, while acknowledging the importance of their cultural and administrative traditions.

These authors identified three country approaches: "those that have made a high commitment to performance management, where the interest is in how they draw up their performance management systems handle the limitations to this approach; countries that have sought to balance performance management with other features and how that plays out in practice; and countries that have sought to selectively draw on performance management techniques while operating within another type of system" (p. 14).

This framework proved to be useful in gathering evidences of several impacts of IS strategies on four major transitions in libraries' evaluation models grouped into four cycles of recognition, reinforcement, uncertainty and weakness (seeTable 2), summarizing a list of dimensions that define Portugal as a country that has sought to balance performance management with other priorities and political and economic influences.

Table 2. IKS landmarks and major transitions v/s main Portuguese performance evaluation models.

	TOPIC		PERSPECTIVE		
	Library	Use	I	E	H
1973		Dialog (online information system)			
TRANSITION I – Cycle of recognition and uncertainty					
1984	*Schwencke Resolution*				
1985	*Resolution Collaboration between libraries in the field of data processing*				
1988	*Plan of Action for Libraries in the European Community (1988–1994)*				
1989		*Information Market Policy Action (1989–1990)*			
1990	*Libraries Programme (1990–1999)*				
1991	*Telematics for Libraries (1991–1998)*				
	LMLN System for collecting and reporting performance information	LMLN System for collecting and reporting performance information			
1992		*Maastricht Agreement*			
1993	PLNN Model of Performance Information Management System	PLNN Model of Performance Information Management System			
1994		*Bangemman Report* *European Commission on Preservation and Access*			
TRANSITION II – Cycle of recognition					
1996	IU-ME Quality Program				
1997		*Digital heritage and Cultural Content* *Culture 2000* *eTen (1997–2006).* *Livro Verde para Sociedade da informação*			
1998		*Information Society Technologies (1998–2002)*			
	Standards	Standards			
1999	MonitorDoc				
2000		*Lisbon Strategy* *CAF* *eEurope 2002* *Culture 2000 (2000–2006)* *eContent (2000–2005)*			
TRANSITION III – Cycle of reinforcement					
2001		TEL – European Library			
	KRCN Performance Evaluation Model	KRCN Performance Evaluation Model			
	KRCN Quality Observation Model	KRCN Quality Observation Model			
2002		*eEurope 2005* *European research area, 2002–2006*			
2003	LMLN Performance Evaluation Integrated System				
2004	ISO 9001 certification				
	LMLN Service Quality Model	LMLN Service Quality Model			
	SIADAP[18] Model				
2005		*i2010* Europeana			
	Mix-model CAF-BSC(-AHP)				
2006	Digital Library Integrated Evaluation Model				
2007		*LIBER-EBLIDA Digitisation Road Map*			
2008	School Libraries Self-assessment Model				
		ENTITLE-LMNL Impact Assessment Framework			
2010		*European Digital Agenda (2010–2020)*			
TRANSITION IV – Cycle of uncertainty and weakness					

Legend: *Perspectives on library performance evaluation:***I***-Internal,***E** *– External,***H** *– Holistic.*

The national and European contexts influence the four transition cycles selected:

3.1. Transition I (1984–1994)

Cycle 1986–1994 – transition to EU performance trends. – We see a phase of recognition of evaluation processes inspired by European libraries. All types of libraries are listed and participate in several plans. The National Library has a particular role as leader of the change process. User needs and satisfaction methods are recognized as management practice and several diagnostic studies of Portuguese libraries are carried out in order to collect information.

One example is the LMLN System for collecting and reporting performance information in public libraries (1991). However, uncertainty and lack of training is assumed by the generality of LIS professionals.

3.2. Transition II (1994–2000)

Cycle 1994–2000 – transition to Information Society national agenda. – This process has its major evidence in the recognition of libraries as partners in the process of building the Information Society in Portugal: public libraries have a strong participation in this phase; governmental libraries try to explore quality models and benchmarking through the Monitordoc self-evaluation model (1999). Quality models like the Common Assessment Framework (CAF) are disseminated mainly after 2000 and many academic and specialized libraries developed successful experiences of self-evaluation.

3.3. Transition III (2000–2010)

Cycle 2000–2005 – transition to e-government policy. – Libraries reinforce their role in the conceptualization and development of performance evaluation models. This emphasis is due to the national policy on public services assessment and the creation of the SIADAP^{+B} model (2004) is one of its sectorial outcomes. This model is intended to be applied in all types of Portuguese libraries. This transition is characterized by the growing interest of academic libraries in quality management and benchmarking.

Cycle 2005–2010 – transition to the national Technological Plan and strong alignment with i2010 strategy – Libraries are considered mainly as information resources for usage. School libraries play an important role in the national pool of performance evaluation of best practices.

3.4. Transition IV (2010-)

Cycle 2010- – transition to a digital agenda in a time of financial crises. – The sector is facing a very relevant challenge. There is a need to re-think the cultural system today in the face of a global and national economical crisis. The role of libraries is uncertain and is dependent upon the performance evaluation orientation provided by national governmental structures. At the same time, opportunities to explore, develop and exchange experiences are scarce. Measuring quality is not as important as before and services are interested in developing approaches centred on resources to prove efficiency. In contrast to previous decades, Excellence is now a new and emerging phenomenon in libraries, not only with organizational impacts but also with a strong emphasis on information society transition moments. This involves an interdisciplinary view (Management, Sociology, Information Science, Communication, and Marketing) that examines the confluence of different variables on evaluation methods and practices. One of those studies is concerned with convergence culture and information environments, considering hybridization as a process of integration and fragmentation, particularly during periods of unusual transition, contingency and negotiability. We believe like **Hobohm (2012**, p. 8) that "we find a concept of quality which indeed stems from the warehouse paradigm and is not rooted in modern customer-oriented quality management".

The construction of the Single Information Market, the Digital Agenda and the Agenda for Culture requires a different understanding of how the new cycles work and how evaluation models can help libraries to survive and advance. Impact metrics are also a new theme related to information management developments. However, due to financial priorities, information-documentation qualifications and skills are less valued in the market and some signs of weakness in the application of evaluation models become progressively more apparent. One of these signs is visible when in governmental libraries managers are chosen from other professional fields, ignoring librarians' performance, leadership and importance. This trend is new in Portugal and reflects a transition to invisibility and loss of career among librarians.

4. DISCUSSION AND FINAL REMARKS

Information Society and library performance evaluation pathways can be seen as being separate areas of interest and separate lines of research. Our meta-evaluation model and frameworks aim at linking and integrating different views and practices over time, as tools to analyse the different cycles of IS public policy transitions and at outlining an agenda for joint research. They do not attempt to cover all the phenomena but, on the other hand, by adopting a longitudinal approach to library performance evaluation, at national or international levels, LIS academics and practitioners gain a better understanding of the present situation, as well as a better chance of making in the future more informed decisions regarding library performance management and evaluation policy. These dimensions provide comprehensive coverage for future multi-dimensional IS constructs concerning library management models, assessment trends and improvement tools. The data collected from this framework are the prerequisite for further modelling and for the implementation gap analysis of performance models.

There is plenty of scope for further study. To complement this examination, we suggest the use of SWOT analysis to clarify and evaluate the importance and relevance of IS developments identified in regard to assessment implementation modalities and PEST analysis to complement the cross-border contexts and evidence-based practices, making visible the multitude of excellences that can be found in multiple evaluation models.

The way forward to achieving the dream of a modern society implies that the library sector highlights the significant performance improvements to create and sustain the evaluation environment that is needed for enabling and encouraging libraries to become best practice organizations.

NOTES

[1]Cost efficiency, cost effectiveness, cost comparison, financial allocation, contingent valuation and value added return on investment.

[2]**L.G. Pinto (2012)** investigated in her doctoral research performance assessment impact on Portuguese Libraries and P. **Ochôa (2012)** investigated in her PhD research professional trajectories in information society transitions. In earlier research, the authors have categorized quality dimensions in Portuguese libraries, which formed the foundation of their current research.

[3]Meta-evaluation is the evaluation of an evaluation or cluster of interconnected evaluations (**Scriven, 1991**).

[4]According to **Wilson (2000)**, "...'user studies' is a term which covers a very wide range of potential research, from the study of users' choices of books from a university library, through reactions to on-line search outputs, to the in-depth analysis of the underlying needs that result in information-seeking".

REFERENCES

1. Aabø, S. (2011). The value of public libraries: a socio-economic analysis. In Belloti, M. (Ed.), Verso un'economia della biblioteca: Finanziamenti, programmazione e valorizzazione in tempo di crisi (pp. 169–176). Milano: Bibliografia

2. Almauk (2012). Economic impact toolkit. Retrieved November 3, 2012, from http://almauk.org/working-together/our_activity/economic-impacts

3. Bonet, L., & Donato, F. (2011). The financial crisis and its impact on the current models of governance and management of the cultural sector in Europe. ENCATC Journal of cultural management and policy, 1, 4–11.

4. Bouckaert, G., & Halligan, J. (2006, September). A framework for comparative analysis of performance management. Paper presented to the Study Group on Productivity and Quality in the Public Sector at the Conference of European Group of Public Administration, Universitá Bocconi, Milan. Retrieved November 13, 2013, from http://webh01.ua.ac.be/pubsector/milan/bouckaert_halligan.pdf

5. Bouckaert, G. (2012). Managing performance: international and national trends. Forum AP, Roma, May 17th. Retrieved September 27, 2012, from http://www.qualitapa.gov.it/

fileadmin/mirror/i-valperf/materiali/2012-05-17_VdP_Slide_Bouckaert.pdf

6. Bradford, S. C. (1948). Documentation. London: Crosby Lockwood

7. Calvert, P. J. (2008). Assessing the effectiveness and quality of libraries. Wellington: Victoria University of Wellington. Retrieved July 7, 2011, from http://researcharchive.vuw.ac.nz /bitstream/handle/10063/1045/thesis.pdf?sequence= 1

8. Chew, K., Stemper, J., Lilyard, C., & Schoenborn, M. (2012) User-defined valued metrics for electronic journals. Library Assessment Conference: Building Effective, Sustainable, practical assessment. In: Proceedings of the 2012 Library Assessment Conference, Charlottesville, VA, Oct 29–31. Retrieved November 13, 2013, from http:// eprints.rclis.org/18727/1/UserDefinedValueBasedMetricsFor ElectronicJournalsPDF.pdf

9. Codagnone, C., & Wimmer, M.A. (2007). Roadmapping eGovernment research: visions and measures towards innovative governments in 2020. Clusone: MY Print. Retrieved November 13, 2013, from http://air.unimi.it/ bitstream/2434/171379/2/FinalBook.pdf

10. Cullen, R. (2005). Operationalising the Focus/Values/Purpose matrix: a tool for libraries to measure their ability to deliver service quality. Presentation at the 6th Northumbria Performance Measurement Conference, August 2005, Northumbria University. Also (2006) Performance Measurement and Metrics, 7 (2), 83–99.

11. Demmke, C. (2006). Governmental, organisational and individual performance. Performance myths, performance 'hype' and real performance. EIPASCOPE, 1, 5–12.

12. ENTITLE (2009). Impact Assessment Framework. Retrieved March 12, 2011, from http://www.entitlelll.eu/eng/Assessment-Framework

13. Gonçalves, M.A., Moreira, B.L., Fox, E.A. & Watson, L.T. (2007). What is a good digital library? A quality model for digital libraries. Information Processing & Management, 43 (5), 1416–1437.

14. Hobohm, H.-C. (2012). Can digital libraries generate knowledge? Contribution to the Cologne Dialogue on Digital Humanities. Retrieved November 13, 2013 from http://www.cceh.uni-koeln.de/files/Hobohm_final.pdf

15. Huijboom, N., & van den Broek, T. (2011). Open data: an international comparison of strategies. European Journal of e-practice, 12 (March/April), 1–13. Retrieved November 13, 2013, from http://www.epractice.eu/files/ European%20Journal% 20epractice%20Volume%2012_1.pdf

16. Imholz, S., & Weil, J. (2007). Worth their weight: an assessment of the evolving field of library valuation. New York: Americans for Libraries Council. Retrieved July 19, 2012, from http://www.ala.org/ala/research/ librarystats/worththeirweight.pdf

17. Markless, S., & Streatfield, D. (2006). Evaluating the impact of your library. London: Facet Publishing

18. Mays, R., Tenopir, C., & Kaufman, P. (2010). Lib-Value: measuring value and return on investment of academic libraries. Research library issues: a bimonthly report from ARL, CNI, and SPARC, No. 271 (August 2010). Retrieved June 2, 2012, from http://works.bepress.com/cgi/viewcontent.cgi?article=1135&context=carol_tenopir

19. Melo, L.B., & Pires, C. (2008). Performance evaluation of academic libraries: implementation model. Paper presented at the 17th Hellenic Conference of Academic Libraries, Ioanina (2008), Retrieved May 2, 2012, from: http://eprints.rclis.org/bitstream/10760/12374/1/b2.Melo_Pires.pdf

20. Nicholson, S. (2004). A conceptual framework for the holistic measurement and cumulative evaluation of library services. Journal of documentation, 60 (2), 164–182.

21. Oakland, J.S. (2003). Total quality management: text with cases. 3rd ed. Oxford: Butterworth-Heinemann

22. Ochôa, P. (2012). Transições profissionais na Sociedade de Informação em Portugal: percursos identitários e ciclos de competências de bibliotecários portugueses (1973–2010). Alcalá de Henares: Universidad de Alcalá. PhD thesis. Retrieved November 13, 2013

23. Ochôa, P., & Pinto, L.G. (2006). Quality - an on-going practice and reflection in a governmental library (1996–2006). In IFLA General Conference and Council, 72, Seul.

24. Ochôa, P., & Pinto, L.G. (2012). Contextos emergentes da excelência. In Congresso Nacional de Bibliotecários, Arquivistas e Documentalistas, 11, Lisboa, 2012, Integração, Acesso e Valor Social: actas. Lisboa: B.A.D. Retrieved July 3, 2013, from http://www.bad.pt/publicacoes/index.php/congressosbad/article/view/299

25. Pinto, L.G. (1998). Portugal. In IFLA Satellite Meeting, Berlim, 1997. Performance measurement and quality management in public libraries: proceedings (pp. 110–119). Berlin: Deutsches Bibliotheksinstitut

26. Pinto, L.G. (2009). Testing the Impact Assessment Framework in Lisbon Municipal Libraries Network. Presentation at the

Budapest Entitle Conference on 'Libraries for Lifelong Learning'. Presentation can be retrieved from http://entitlelll.eu/hun/Meetings/Final-Conference/Presentations/Testing-the-Impact-Assessment-Framework-Lisbon

27. Pinto, L.G. (2012) – Dimensões estratégicas e emergentes da avaliação do desempenho: modelos, dinâmicas e impactos na Rede de Bibliotecas Municipais de Lisboa (1989–2009). Alcalá de Henares: Universidad de Alcalá. PhD thesis.

28. Pinto, L.G., Ochôa, P., & Vinagre, M.H. (2009). Integrated approach to the evaluation of digital libraries: an emerging strategy for managing resources, capabilities and results. In M. Heaney (Ed.), Library statistics for the 21st century world (pp. 273–288). München: K. G. Saur

29. Plum, T., Franklin, B.R., Kyrillidou, M., Roebuck, G., & Davis, M. (2010). Measuring the impact of networked electronic resources: developing an assessment infrastructure for libraries, state, and other types of consortia. Performance measurement and Metrics, 11 (2), 184–198.

30. Poll, R. (2007). Benchmarking with quality indicators: national projects. Performance Measurement and Metrics, 8 (1), 41–53.

31. Poll, R. (2012). Can we quantify the library's influence? Creating an ISO standard for impact assessment. Performance Measurement and Metrics, 13 (2), 121–130.

32. Portugal. Ministério da Educação e Ciência. Gabinete da Rede Bibliotecas Escolares (2011). Modelo de avaliação da biblioteca escolar. Lisboa: Gabinete RBE. Retrieved September 19, 2011, from http://www.rbe.min-edu.pt/np4/83.html

33. Rankin, C. (2012). The potential of generic social outcomes in promoting the positive impact of the public library. Evidence

from the National Year of Reading in Yorkshire. Evidence Based Library & Information Practice, 7 (1), 7–21.

34. Scriven, M. (1991). Evaluation thesaurus, London: Sage

35. Shi, X., & Levy, S. (2005). A theory-guided approach to library services assessment. College & research libraries, 66 (3), 266–277.

36. Staes, P., & Thijs, N. (Eds.) (2010). Growing Towards Excellence in the European Public Sector: A decade of European collaboration with CAF. Maastricht: EIPA. Retrieved November 13, 2013

37. Streatfield, D., & Markless, S. (2009). What is impact assessment and why is it important?. Performance measurement and metrics, 10 (2), 134–141.

38. Tanner, S., & Deegan, M. (2012). Inspiring research, inspiring scholarship. London: JISC.

39. Tenopir, C. (2012) Beyond usage: measuring library outcomes and value. Library Management, 33 (1/2), 5–13.

40. Town, S. (2011). Value, impact and the transcendent library: progress and pressures in performance measurement and evaluation. Library Quarterly, 81 (1), 111–125.

41. Van Dooren, W. (2005). What makes organisations measure? Hypotheses on the causes and conditions for performance measurement. Financial accountability & management, 21 (3), 363–383.

42. Van Dooren, W. (2011). Why is performance information not used? In search for some common barriers. Paper presented at the 2011 ECPR general Conference (Reykjavik), University of Antwerp, Belgium

43. Van Dooren, W., Bouckaert, G., & Halligan, J. (2010). Performance management in the public sector. London: Routledge

44. Vinagre, M.H., Pinto, L.G., & Neves, J. (2011). Giving voice to children and young adults: a methodological approach to the evaluation of service quality in libraries. In Northumbria International Conference on Performance Measurement in Libraries and Information Services (PM8), 8, Florence (2009). Libraries plus: adding value in the cultural community: proceedings (340–358 pp.). Newcastle upon Tyne: Northumbria University Press

45. White, L.N. (2002). Does counting count: an evaluative study of the use and impact of performance measurement in Florida public libraries. Tallahassee: Florida State University

46. Wilson, F., & Town, J.S. (2006). Benchmarking and library quality maturity. Performance Measurement and Metrics, 7 (2), 75–82.

47. Wilson, T.D. (2000). Recent trends in user studies: action research and qualitative methods. Information research, 5 (3).

48. Winkworth, I., & Gannon-Leary, P. (2000). Library performance measures: government perspectives. In Northumbria International Conference on Performance Measurement in Libraries and Information Services, 3, Northumberland, 1999 – Value and impact: proceedings (pp. 61–65). Newcastle: Information North

9

CHAPTER

A Global Approach to Digital Library Evaluation

Giuseppina Vullo

Humanities Advanced Technology and Information Institute (HATII), University of Glasgow

ABSTRACT

This paper describes the key advances on digital library evaluation research. The paper provides a comparison of the existing models, the current research questions in this area, an integrated LIS-oriented evaluation framework, and a selection of international projects.

Keywords: digital libraries, digital library evaluation, digital library evaluation models, quality

INTRODUCTION

More and more efforts have been made in the last decade to evaluate digital libraries and to build global evaluation models, even if an accepted methodology that encompasses all the approaches does not exist. Research and professional communities have specific viewpoints on what digital libraries are, and they use different approaches to evaluate them. Evaluating digital libraries is a challenging activity, as digital libraries are complex, dynamic and synchronic entities which need flexible approaches.

Since 1999, when Christine Borgman described the gap between the perspectives of digital library researchers and professionals (Borgman, 1999), several initiatives have been undertaken to establish a framework for exchange between the two communities. The reference definitions of 'digital library' in this work are:

1. 'Digital libraries are organizations that provide the resources, including the specialized staff, to select, structure, offer intellectual access to, interpret, distribute, preserve the integrity of, and ensure the persistence over time of collections of digital works so that they are readily and economically available for use by a defined community or set of communities' (Waters, 1998), which has been adopted by the Digital Libraries Federation in 2002.

2. 'A possibly virtual organization that comprehensively collects, manages, and preserves for the long term rich digital content, and offers to its user communities specialized functionality on that content, of measurable quality and according to codified policies', formulated within the DELOS project (Candela et al., 2008).

We will also keep a background library science approach taking into account Ranganathan's five laws (Ranganathan, 1931):

1. Books are for use.

2. Every reader his [or her] book.

3. Every book its reader.
4. Save the time of the Reader.
5. The library is a growing organism.

State-of-the-art

Digital library (subsequently DL) evaluation has been investigated since the end of the 1990s, when Saracevic and Kantor (Saracevic & Kantor, 1997) reviewed the traditional libraries' evaluation criteria identified by Lancaster (Lancaster, 1993), and Saracevic (Saracevic, 2000) systematised the issue within a continuative approach, highlighting the need to focus on the DL mission and objectives. According to Saracevic, considering evaluation as the appraisal of the performance or functioning of a system, or part thereof, in relation to some objective(s), the performance can be evaluated as to: effectiveness (how well does a system do what it was designed for?)

- efficiency (at what cost, in terms of money or time?)
- a combination of these two (i.e. cost-effectiveness) (Saracevic, 2000, p. 359).

Saracevic also indicated two evaluation levels, which is challenging to integrate:

- user-centred level (which can be social, institutional, individual or focused on the interface)
- system-centred level (which can be focused on engineering, processing or content) (Saracevic, 2000, p. 363–364).

In the same year, Marchionini (Marchionini, 2000) proposed the application of the same techniques and indicators used for traditional libraries, such as circulation, creation and growth of collections, user data, user satisfaction, and financial stability indicators. According to Marchionini, a DL evaluation can have different aims, from the understanding of basic phenomena (e.g. the users' behaviour towards IR tools) to the effective evaluation of a specific object. He presented the results of a longitudinal analysis of the Perseus DL, which lasted more than ten years (Marchionini, 2000). Among the evaluation corollaries of that study, he stated that successful DLs should have:

1. clear missions
2. strong leadership and a strong talent pool
3. good technical vision and decisions
4. quality content and data management
5. giving users multiple access alternatives
6. ongoing evaluation effort (Marchionini, 2000).

Successively, some guidelines to evaluate DLs were proposed (Reeves, Apedoe & Woo, 2003), focusing on the decision process that is behind any evaluation; and the need to focus on the 'global' impact that a DL has on its users and on society was highlighted (Chowdhury & Chowdhury, 2003) by integrating LIS, IR and HCI criteria. Through the analysis of eighty DL case studies, Saracevic observed the small quantity of 'real data' in comparison to the explosion of meta-literature (Saracevic, 2004), concluding that there is no 'best' methodology: different aims can lead to heterogeneous methods.

The development of an evaluation model has been carried forward by

the DELOS project; its evaluation schema initially had three dimensions:

1. data/collection
2. system/technology
3. users/uses (Fuhr et al., 2001).

This schema was then integrated into Saracevic's four evaluation categories (Saracevic, 2004; Fuhr et al., 2007).

A quality model for digital libraries was elaborated in 2007 within the 5S (Streams, Structures, Spaces, Scenarios, and Societies) theoretical framework (Gonçalves et al., 2004; Gonçalves et al., 2007); the model was addressed to digital library managers, designers and system developers, and defined a number of dimensions and metrics which were illustrated with real case studies.

Within the DELOS Digital Library Reference Model (Candela et al., 2008), quality facets and parameters have been investigated to model the Quality domain. Quality is defined as 'the parameters that can be used to characterise and evaluate the content and behaviour of a DL. Some of these parameters are objective in nature and can be measured automatically, whereas others are inherently subjective and can only be measured through user evaluation (e.g. in focus groups' (Candela et al., 2008, p. 20).

The ongoing EU-funded project DL.org is currently investigating and identifying solutions for interoperability, according to the six domains of the DELOS Reference Model (Architecture, Content, Functionality, Policy, Quality and Users) (Candela et al., 2008), and its Quality

Working Group aims to continuing the research on quality parameters and dimensions developed within DELOS.

Research advances on digital library evaluation and quality are especially needed considering the amount of national and international collaborative projects aiming the interoperation between diverse DLs, and their connection with individuals, groups, institutions, and societies; they can also have a crucial role within DL projects political and social acceptance.

Digital Library Evaluation Models

Concepts and models for evaluating digital libraries come mainly from three research areas: library and information science (LIS) studies, computer science studies, and human-computer interaction (HCI) studies.

They can adopt the following types of approach:

1. content-based approach (DLs as collections of data and metadata);
2. technical-based approach (DLs as software systems);
3. service-based approach (DLs as organisations providing a set of intangible goods, i.e. benefits);
4. user-based approach (DLs as personal and social environments).

Considering the dynamic nature of digital libraries and the spread of projects dedicated to them, the research on comprehensive evaluation models is quite limited. In this study we present a selection of models that not always come from the DL field; however, they are considered relevant for building global DL evaluation frameworks.

The first comprehensive and multidimensional model comes from information systems research. It is known as the D&M IS success model (DeLone & McLean, 1992) and was updated in 2003 (DeLone & McLean, 2003), when the authors decided to revise it considering the advent and explosive growth of e-Commerce. The updates concerned the adding of a 'service quality' measure as a new dimension, and the grouping of all the 'impact' measures into a single impact or benefit category called 'net benefit' (in the original model, there were 'individual impact' and 'organisational impact'). The D&M IS success model identifies the interrelationships between six variable categories involved in the 'success' of information systems: 'information quality', 'system quality', 'service quality', 'intention to use/use', 'user satisfaction', 'net benefits'.

The three types of quality (which — in the DL world — correspond to content quality, DL system quality and DL services quality) concur to the quantitative and qualitative interactions with the information system, respectively 'intention to use/use' and 'user satisfaction'. The final entity of the model is 'net benefits', which within the DL world could become 'social benefits', either to the individuals or to groups and communities. This model does not offer any specific quality parameter or metrics: on the contrary, it aims to be simple and as general and applicable as possible.

The second model comes for the Library Science field and was created in 2004 for the holistic evaluation of traditional library services (Nicholson, 2004). The model is a pyramid describing the evaluation workflow; it adopts an operational approach, and identifies core steps and actors involved in the evaluation process. Nicholson's model is relevant not only because it is the first one that aims to consider libraries' evaluation holistically, but mostly because it takes into account the role of the administrators — who are also the decision-makers — at the head of the evaluation pyramid. At the basis of Nicholson's pyramid there is the measurement matrix, i.e. the measurements from different topics and perspectives; the upper level is constituted by the evaluation criteria; the highest level corresponds to the evaluation viewpoints which are classified hierarchically (from the lowest to the highest) as evaluations by 'users', by 'library personnel', and by 'decision-makers'. Once the decisions have been made, the evaluation cycle moves from the pyramid's top to its bottom, with the 'changes implemented by the library personnel', 'users impacted by changes', 'evaluation criteria selected to measure impact on system' and, at the basis the pyramid again, the 'measurements from different topics and perspectives selected', where the cycle starts again (Nicholson, 2004). Nicholson's focus is the organisational context of evaluation, and it does not explain how the different viewpoints and measurements can be combined or integrated, nor the quality parameter and metrics involved.

The third model — known as 'a generalised schema for a digital library' — is the result of research developed within the EU-funded DELOS project. It constitutes the first holistic model specifically created for DL evaluation from the research community (Fuhr et al., 2001). The model describes the DL domain and its three core entities — 'system/technology', 'data/collection', and 'users' — all directing to a fourth entity called 'usage'. The DL domain is over-arched by a larger circle called 'research domain'. The research domain identifies the research areas involved in the four entities of the DL domain as follows:

1. system/technology: system and technology researchers;
2. data/collection: librarians, LIS researchers;
3. users: publishers, sociology of science, communication researchers;
4. usage: HCI, librarians, systems researchers (Fuhr et al., 2001).

This model effectively illustrates the heterogeneity of research fields involved in DLs. However — excluding policy makers, managers, senior librarians and administrators — does not take into account the organisational context of the DL.

The fourth and most recent model, grounded on Saracevic's evaluation dimensions, identifies the most important evaluation criteria according to the different stakeholders of digital libraries, within a holistic perspective ('Holistic DL evaluation model') (Zhang, 2010). Zhang identified the most relevant digital library evaluation criteria among five groups of survey participants (administrators, developers, librarians, researchers, users), according to Saracevic's six dimensions (content, technology, interface, service, user, context).

While conducting a comparative study on digital library evaluation models, I developed a LIS-oriented framework, as illustrated in Figure 1. The model includes both the user and system perspective. The arrows indicate the routes corresponding to the two perspectives, i.e. the DL core directions, respectively 'use' and 'mission'.

The core entities of the DL are organisation, content, services and users. According to these four entities the evaluation can focus on organisational aspects (such as management and policies), content aspects (quality of data, metadata, digital collections and even digital libraries), service aspects (quality of technologies or quality of design), or users' aspects (quality of interactions between users and the DL).

Figure 1. A LIS-oriented framework for DL evaluation.

Developing metrics: EQUINOX, the DLib Test Suite and DigiQUAL

Among the first European projects on digital library evaluation, EQUINOX — Library Performance Measurement and Quality Management System, run from 1998 to 2000 — was the first one within which a set of specific performance indicators were developed. The intention of this set of indicators was to enhance the indicators for traditional library services presented in ISO 11620: Library Performance Indicators. The project delivered fourteen performance indicators, a consolidated list of datasets, a glossary and a list of collection methodologies which are still publicly available on the project website.

During the same years, the D-Lib Working Group on Digital Library Metricsran 'The DLib Test Suite' (Larsen, 2002), which was an 'early attempt at organizing a rigorous and well supported test bed to enable comparative evaluation of digital library technologies and capabilities' (Larsen, 2002), which was sponsored by the DLib Forum. The DLib Working Group elaborated a standard set of data for quantitative and comparative research, to evaluate and compare the effectiveness of digital libraries and component technologies in a distributed environment (Larsen, 2002).

Regardless of the efforts of EQUINOX and the DLib Metrics Working Group, there is a lack of common strategies for digital libraries evaluation. Within the ARL (Association of Research Libraries) an operational evaluation protocol has been developed, called DigiQUAL. DigiQUAL aims to provide a standard methodology to measure DL service quality and is grounded on the LibQUALprotocol, which is used to measure service quality by traditional libraries. DigiQUAL has identified more than one hundred eighty items around twelve themes related to digital library service quality (Lincoln, Cook & Kyrillidou, 2004).

CONCLUSIONS

In this paper we indicated interdisciplinary routes towards a global approach to DL evaluation, describing the state-of-the art, proposing models from different research fields, presenting a LIS-oriented framework and selecting the most relevant projects in this area. There is no common agreement on how to evaluate DLs and evaluation activities are still low-prioritised issues within the DL field. Several assessment methodologies have been built and the interdisciplinary research is growing, while a broadly accepted model is still lacking.

REFERENCES

1. Borgman, C.L. (1999): 'What are digital libraries? Competing visions', *InformationProcessing and Management*, 35(3), 227–243.

2. Candela, L. et al. (2007): 'Setting the foundation of digital libraries', D-Lib Magazine, 13, 2007.

3. Candela, L. et al. (2008): The DELOS Digital Library Reference Model. Foundations for Digital Libraries, Version 0.98, Project no. 507618, DELOS,

4. Chowdhury, G.G. and S. Chowdhury (2003): Introduction to Digital Libraries. Facet Publishing, London.

5. DeLone, W.H. and E.R. McLean (2003): 'The DeLone and McLean model of information systems success: a ten-year update', J. Management Information Systems, 19(4), 9–30.

6. Fuhr, N. et al. (2001): 'Digital libraries: a generic classification and evaluation scheme', Proceedings of ECDL 2001. LNCS, 2163, 187–199. Springer, Heidelberg.

7. Fuhr, N. et al. (2007): 'Evaluation of digital libraries', Int. J. Digital Libraries, 8(1), 21–38.

8. Gonçalves, M.A. et al. (2004): 'Streams, Structures, Spaces, Scenarios, Societies (5s): A formal model for digital libraries', ACM Trans. Inf. Syst., 22(2), 270–312.

9. Gonçalves, M.A. et al. (2007): '"What is a good digital library?" A quality model for digital libraries', Information Processing and Management, 43(5), 1416–1437

10. Lancaster, F.W. (1993): If you want to evaluate your library, 2nd edn. Morgan Kaufman, San Francisco.

11. Larsen, R.L. (2002): The DLib Test Suite and Metrics Working Group. Harvesting the Experience from the Digital Library Initiative, University of Maryland, publicly available at.

12. Lincoln, Y.S., C. Cook and M. Kyrillidou (2004): 'Evaluating the NSF National Science Digital Library Collections: Categories and Themes from MERLOT and DLESE', paper presented at the MERLOT International Conference, Costa Mesa, California.

13. Marchionini, G. (2000): 'Evaluating digital libraries: a longitudinal and multifaceted view', Library Trends, 49(2), 304–333.

14. Nicholson, S. (2004): 'A conceptual framework for the holistic measurement and cumulative evaluation of library services', J. Documentation, 60(2), 164–182,

15. Ranganathan, S. (1931): The Five Laws of Library Science. Edward Goldston, London

16. Reeves, T.C. X. Apedoe and Y.H. Woo (2003): Evaluating Digital Libraries: A UserFriendly Guide. National Science Digital Library, University of Georgia, Athens, GA.

17. Saracevic, T. and P.B. Kantor (1997): 'Studying the value of library and information services. Part II. Methodology and taxonomy', J. American Society for Information Science, 48(6), 543–563.

18. Saracevic, T. (2000): 'Digital library evaluation: toward an evolution of concepts', Library Trends, 49(3), 350–369.

19. Saracevic, T. (2004): 'Evaluation of digital libraries: an overview', paper presented at the DELOS Workshop on the Evaluation of Digital Libraries,

20. Su, L.T. (1992): 'Evaluation measures for interactive information retrieval', Information Processing and Management, 28(4), 503–516.

21. Waters, D.J. (1998): 'What are digital libraries?', CLIR Issues, 4, 1998,

22. Zhang, Y. (2010): 'Developing a holistic model for digital library evaluation', J. American Society for Information Science, 61(1), 88–110.

10
CHAPTER

"Just How Are We Going to Satisfy Our Research Customers"

Sir Brian K. Follett

INTRODUCTION

The challenge is how to provide our researchers with ready and unimpeded access to all types of research information, whether it is held on paper or electronically. This has been, of course, the role of research libraries for hundreds of years so why is it a particular challenge at the present time? May I suggest that the problems are acute because a number of factors are all occurring at the same time. Let me mention some of these.

THE FACTORS DRIVING CHANGE

1. Governments now appreciate that „research and its development into commercial practice" are key factors in future national prosperity. This is a change from a decade ago and it means they are investing more heavily in:

- Increasing the overall amount of research, especially in the new biology. • Stimulating industrial research and development by various tax breaks.

- Stimulating scientists to develop their inventions into patents and spin out companies.

- Investing in the infrastructure. It is especially here that we come into play because in my view libraries are as important a part of the infrastructure as a new synchrotron or molecular biology facilities.

2. There is continuing growth in the volume of research being published and this alone places libraries under strain. Even the largest libraries – say the British Library or Harvard – can no longer collect all research materials. The situation is worse if one analyses it in detail because in many respects universities are small enterprises with all the advantages culturally and educationally but all the weaknesses when it comes to providing major infrastructure facilities like a library. Looking at a few statistics rather emphasises the point:

Harvard has 14 million volumes on its shelves, adds 300,000 annually, subscribes to 100,000 serials and spends $70 million annually. A typical „big ten" school (e.g. Wisconsin) has 5 million volumes, adds 150,000 each year, subscribes to 40,000 periodicals and spends $30 million annually. In the UK a strong civic university has 2 million volumes, adds 30,000 annually, subscribes to 10,000 serials and spends $6 million each year. Even the great Bodleian Library at Oxford would only appear just in the top US university library statistics.

This drives home two points. Firstly, that at least in the UK our university libraries are small (bluntly, we may have too many universities if one looks at matters from the research viewpoint) and, secondly, they are not well funded. This shows up especially in their ability to purchase serials. World-class universities purchase at least 40,000 serials: in Britain only Cambridge and Oxford achieve this number, all the others including very serious research universities indeed rarely exceed 10,000. My conclusion is that in the absence of financial largesse from each university – say a tripling in the periodicals budget for a beginning – then alternative solutions have to be found if we are to match our cousins across the atlantic.

3. Inflation of periodicals has outstripped normal domestic inflation by a factor of three. Part of this has come from increased volumes being published but part of it has arisen because, frankly, journal publishing is a highly profitable business. Academics must publish their material and indeed their careers hang directly upon the quality and sometimes the quantity of their published work. The end result is a seller's market! Libraries are invariably in the public sector and so have much less possibility of imposing market forces, upon their users (e.g. charging per visit). Government does not see this and so they apply inexorable downward pressure upon budgets of the parent organisations.

4. Finally, there is the Worldwide Web and IT! There is no doubt about the transforming nature of this technology and articles are being written about it daily. In our context the most significant changes may well relate to the fact that nowadays anyone can publish anything anywhere, and secondly that material can be accessed directly so bypassing the library completely. Either or both of these changes could alter the research library beyond all recognition and if commercial providers stepped into this market we could see many functions of the research library disappear. So far this has not happened and the sheer volume of new research material being added daily has precluded the development of a robust business model. However, deep trends are obvious and the research user will certainly go to the most convenient source of

information: he or she has no especial loyalty to their university library!

THE CENTRAL DESIRES OF THE RESEARCHER

One thing remains true, however, and it may prove to be the most important and durable of all: quality assurance of the research product. If one asks young faculty about the future they see an electronic world. They wish for journal content to be available at the desktop and they wish for far more sophisticated search engines to be available so that their particular research interest can be sifted accurately without throwing up 6,000 journal articles a week! These desires apply especially in the sciences and the emphasis is different in the humanities where „the jury is still out" on the future of the scholarly monograph and the degree to which it will migrate to the electronic only version. However, the humanities scholars seek something else from the electronic world. They wish to see substantial amounts of primary research material made available digitally so that undertaking the actual research via the web rather than by visiting libraries becomes a real possibility. Incidentally this approach to primary research may begin to apply in the sciences also where huge data sets are starting to be produced electronically in areas as diverse as weather forecasting, high energy particle physics, gene sequences and microarray sets of gene expressions information. Nevertheless their primary desire is for assurance about quality. That is good because society must never reach a point where the veracity of scholarly journal articles is diminished.

THE BROAD SOLUTION

1. We cannot predict precisely how the commercial and scientific factors will play out but we should hold fast to our role of ensuring only high quality materials are stored in our libraries.

2. The model of the future we should try to ensure is encapsulated in the latest buzz word: hybrid library. In effect the library becomes a portal through which the researcher enters. They are then offered a range of products and powerful search engines which allow them to move between the products and decide what they need hopefully – at least for me – it will involve them in undertaking a trip to the library to sit and ponder a book but it is rather more likely to involve them accessing an electronic version of a journal article. Already at my old university around 50% of all the periodicals are being accessed electronically. Separately, they might access other electronic library materials held in different uni-versities and given the mark of respectability by being accessed through one's own library, or a national collection of materials assembled from across the world. They might also access backruns of journals which have been digitised: the JSTOR re-publisher model. Finally, and in my view critically, our researcher may wish to access the primary data sets held nationally or internationally. Our role as a research library is to serve this range of demands. It sounds difficult but honestly it is what most of you are doing already

3. All across the world this model is being developed and refined as we speak but many of us think something at a „higher" level is also required since success will only come about if we integrate these functions together.

RESEARCH LIBRARY SUPPORT GROUP

Across the world we are watching collaborative developments emerge. They are developing for a host of reasons: because of extreme distances between libraries in centres of small population (e.g. Australia), because a nation is trying to leap over the stage of having multiple research libraries and goes for a nationwide solution (e.g. China), or because a particular region wishes to maximise its already powerful resources (e.g. The Midwest Consortium of the Big Ten universities plus Chicago).

Here in the UK we have spent some time in the last decade encouraging most strongly the working together of our university libraries. This started from a report eight years ago in which some of us in this room were involved and it led to increasing the visibility of the university's library (an important political point: libraries are embedded in larger institutions), the E-lib initiative, the creation of a national distributed electronic collection (DNER), the decision to fund certain libraries in order for them to maintain a free open door policy towards researchers from other universities, and to that all important feature, preparing the ground for further changes in this decade. So far these efforts have been viewed by the universities and by the libraries as successful and they have encouraged us to think about a possible next evolutionary step.

That step is to trigger much more significant and deeper working together between university research libraries which will, let me say, tread on some hallowed librarian's ground, and to do all this in partnership with the British Library and the two smaller national libraries in Aberystwyth and Edinburgh. Bringing together the national libraries with the university libraries is a significant step (and one for which we can thank Lynne Brindley who has crossed the floor from academe to the british library).

The end result has been a decision to create what we are calling „the Research Libraries Support Group". Guess who is to chair this beast! In addition it contains some of the usual suspects: Lynne herself, Reg Carr from the Bodleian and Michael Anderson from Edinburgh University. We are truly an „official" committee created by the four UK university funding councils along with the British Library. Our terms of reference and vision make it apparent what we have to do in practical terms:

1. We are charged with formulating a national strategy to ensure uk researchers have ready and unimpeded access to all types of research information, whether held on paper or electronically.

2. Put simply we should strive to develop a national model which results in the best collaborative research library arrangements possible, and to think of financial drivers which will cause this to happen.

3. This probably means developing the „hybrid library" concept so that it can apply across the country.

4. To address directly the long-term management and development of research collections through deep collaboration. This takes me back to my earlier point about the relatively small size of UK libraries in terms of serials and book purchases. We know that bottom-up collaborations will work but we suspect the pace of change is too slow.

5. To think about and develop further the „national electronic library" of which the DNER is a strong manifestation.

6. To seek a national strategy for digitising existing collections of primary research material.

7. To seek a national solution to the preservation of the so-called legacy collections.

8. To advise on where we find and develop the critical underpinning requirements which might range from catalogues to the search engines in the back office of our hybrid libraries.

9. Finally, not to rediscover the wheel and, therefore, to link our efforts with what is happening across the world with our friends and colleagues elsewhere

In many respects it is slightly daunting to set it out in this fashion, especially as the funding councils would like a report in the spring of next year. However, it is very exciting and I am optimistic that whilst we shall doubtless fall short of perfection we may make another step change in the ways in which our libraries work together. What is more any solutions we find will have an applicability across the world and so we might make a contribution to the objective which is to serve our research users well so that they can advance our understanding of this remarkable universe and planet we live in.

11

CHAPTER

Information Handling Iin Collaborative Research

Ellen Collins, Michael Jubb

Research Information Network

ABSTRACT

UK public policy makers have a growing interest in collaborative research, where academics work with public, private or third sector partners on a joint project which supports the partner's aims. This paper reports on the findings of five case studies, looking at how information is sourced, managed, used and shared within collaborative research projects. It finds that researchers within collaborative projects have similar information management issues as are known to exist within academia more broadly, but that the specific conditions which govern research collaborations mean that interventions to improve or support information management must be carefully tailored.

Key words: knowledge economy, collaboration, information handling

1. INTRODUCTION

All research is inherently a collaborative enterprise, building upon the expertise of colleagues, predecessors and competitors to arrive at new discoveries and understandings. In recent years, however, 'collaboration' has taken on a new meaning in UK public policy as it relates to the research sector. Increasingly, the term has come to refer to the relationships and projects developed between academic researchers and their counterparts in the public, private and third sectors. A key feature of these relationships is that they should be targeted towards a specific outcome, solving problems in order to benefit industry, society and the economy (Lambert, 2003). Collaborative research is one element of a wider spectrum of interactions between higher education institutions (HEIs) and external organisations, including consultancy, education and training, contract research and intellectual property deals: all of these have become increasingly important aspects of UK higher education funding.

The wider context of UK higher education policy has reflected and supported this drive for some time. During the 1990s, government began to encourage universities to become more proactive about encouraging the commercialisation of their research in-house and through collaborations with external organisations. A series of Higher Education Funding Council for England (HEFCE) interventions established funds to support such work. The Lambert Review and the follow-up Science and Innovation Investment Framework demonstrated the Labour government's continued interest in this area, as did the departmental reshuffle in 2007 which brought higher education and business into a single portfolio for the first time, signalling a clear agenda for academia by separating it from primary and secondary education (Lambert, 2003). The Sainsbury Review and subsequent Innovation Nation white paper focused on the business demand side, while a series of concurrent reviews looked at the skills that graduates needed to meet employers' business demands (Sainsbury, 2007). The 2011 Innovation and Research Strategy for Growth (BIS, 2011) indicates the Coalition Government's continued interest in this area, as does the Wilson Review, which reported in early 2012 (Wilson, 2012).

The long history of political support has borne fruit, in the form of a growing number of relationships between research establishments — HEIs and others — and public, private and third sector partners. The most recent data from HEFCE indicates a 4% growth in knowledge exchange between 2008–9 and 2009–10 among UK HEIs, and a growth in income from such sources by 35% from 2003–4 to 2009–10. Collaborative research income increased by just over 2% between 2008–9 and 2009–10, although changes to the survey methodology mean that this figure is not entirely reliable. The report also highlights the growing importance of non-commercial collaborators, such as government bodies or third sector charitable organisations, as partners in these enterprises. (HEFCE, 2011). A separate survey, undertaken by the Department for Business, Innovation and Skills (BIS), looks at public sector research establishments (PSREs) and their research interactions with external organisations. The most recent edition of this survey found that 25,000 new collaborative research projects began in 2007–8, and that 93% of PSREs were doing collaborative research (BIS, 2009).

So collaborations, as one possible interaction between research establishments and external bodies, have become increasingly important to universities and their finances. The question then arises: how should university librarians ensure that they engage with collaborative projects and contribute to successful research outcomes? To understand this, we must first understand the specific circumstances which govern collaborative research.

Previous studies have shown that a researcher's environment and collaborators can strongly affect their information behaviour, but that they also retain strong associations and preferences from their own, early, training (RIN/NESTA, 2010; Bulger et al., 2011; Meyer et al., 2011). The cultures in business and higher education are recognised to be rather different, and the relationships can be difficult to manage (Lambert, 2003).

There are several stages of information handling for research projects, each of which offers challenges for researchers used to working in different environments. A new research project involves sourcing information that has already been published on the subject. Researchers outside academia generally have worse access to published scholarly information than those within academic paywalls, and employ different coping strategies when they cannot access the information that they need (CIBER, 2011). At this early stage, researchers will also often seek to discover and share unpublished information which may have a bearing on their research project. But researchers are often very protective of unpublished research and data: in academia, because they wish to protect future publication and funding opportunities, and in industry because of the need to protect commercial interests. Furthermore, there is some confusion and concern within industry about Freedom of Information legislation, and their vulnerability under such legislation if they share confidential data with an HEI. During the research process, new information is discovered and must be shared between different parts of the research team. Although there are standards and protocols for data collection and sharing in many disciplines, at an individual level researchers' behaviour can be idiosyncratic, and there is no guarantee that their techniques will translate into a more commercial research environment (or, indeed, that the commercial techniques will make sense in academic settings). Finally, academic researchers want to publish and share their findings, as the chief way of securing recognition and credit for their work. In many cases, this may actively undermine more commercial strategies, where there are uncertainties about the ramifications of such publication for copyright and intellectual property (CIBER, 2010). And non-commercial partners, such as public sector or charitable organisations, may want to publish findings more rapidly to reach a wide community of stakeholders, thereby diminishing the academics' chances of having an article accepted in a highly-rated journal.

The circumstances governing information handling in collaborative research are thus highly complex and nuanced. If librarians and information professionals are to engage successfully with such projects, they must understand exactly how these circumstances affect the course of individual research projects. To promote such understanding, the Research Information Network commissioned a set of case studies, which looked at five collaborative research projects to identify some common themes and issues, and some possible actions for library staff in engaging with such projects (Jordan et al., 2011). This paper considers the findings of the study

2. METHODS

Five cases were selected, each of which included at least one university and one external partner. Overall, the cases were selected to reflect a range of criteria which we considered likely to affect information handling behaviour: these included the discipline of the collaboration, the number and types of partners involved, the stage of the collaboration — from early-stage to completed — the scale of the project and the types of external support received. Table 1 provides a brief overview of the five cases. Within each case, initial interviews with project managers helped establish the broad structure of the collaboration. Subsequent interviews with key individuals, identified in the initial stage, provided greater insight into the information management process within the collaboration. Finally, we undertook a series of workshops across the five collaborations to identify recommendations for future best practice.

Table 1: Participant cases.

	Discipline	Aim	Partners
Intelligent Care	Technology	Developing a monitoring system for older people living in residential dwellings	1 commercial partner 1 university
UK Reactics	Engineering	Exploring the potential commercial applications of novel auxetic materials (materials which expand when stretched)	6 commercial partners 3 universities
PRISMA	Health / Social care	European priorities for research and measurement in end-of-life care	11 partners, including 4 universities, with a range of public and third sector bodies
Structural Genomics Consortium (Oxford)	Structural biology	Pre-competitive mapping of protein structures to support new drug discovery	11 funders – mix of public and charitable funding and pharmaceutical companies 3 universities
Locating Communications Heritage	Arts / Digital technology	Developing a platform which enables mobile phone users to access history of communications and IT	1 university 1 third sector (museum) 2 commercial

Source: Jordan et al. (2011), Information handling in collaborative research. Research Information Network, p.13.

It is important to note that these cases represent nothing more than a snapshot of a few research collaborations. It would be unwise to draw far-reaching conclusions on the basis of such evidence. Nonetheless, we are confident that they represent a useful insight into researcher behaviour, which can be used to support investigations and interventions by library and information professionals in universities.

3. THE CHALLENGES FOR INFORMATION HANDLING

The research uncovered a number of challenges for effective information handling in research collaborations. Many of these will be familiar to library and information professionals, who already encounter indifference and ineffectiveness among many of their academic researchers. Others relate more closely to the specific conditions within research collaborations.

The way that information is handled in collaborative research depends upon a number of factors. Perhaps most important among these is the primary purpose of the collaboration, and the nature of the area that it is investigating. Unsurprisingly, a collaboration which is focused upon creating a commercial product handles information differently from one which hopes to influence practice and behaviour among a user community. The effect of these differences is most readily observed around the dissemination of information. The Intelligent Care project focused upon developing new technologies which were not patentable; therefore, academic researchers on the project were very limited in their ability to publish early-stage research, as there was nothing to prevent the company's competitors using their findings to bring an alternative product to the market. UK

Reactics, while also focused upon a commercial outcome, was producing research that was more likely to lead to a patent and so publication of early results was more acceptable. Although the commercial partners had 28 days to decide whether to veto a proposed publication in order to protect their intellectual property, only one publication was significantly delayed under these terms. In contrast to these two commercial projects, PRISMA sought to disseminate its findings as widely as possible, and in this case the pre-project agreement actually protected the academics, ensuring that they were able to publish findings before they were distributed among PRISMA's stakeholder community. The SGC represented a very interesting case, in that it required a high level of openness in order to protect the long-term commercial aims of the collaboration: in order to secure input from rival pharmaceutical companies, the researchers had to guarantee that no company would gain a competitive advantage by seeing the findings before any others, and the simplest way to do this was to release them to the funders, the companies and the world at large at exactly the same time. Information handling was also visibly affected by project purpose at an earlier stage in the collaboration: researchers working on Intelligent Care, for example, had to spend time investigating the market for a product rather than focusing exclusively on the academic literature.

Established cultures and norms were also important in defining information behaviour; in our five cases, there were noticeable differences between academics and their non-academic partners. For example, within academia there was a prevailing assumption of openness — researchers would share information not only with their own team but also with teams outside their own institution, even where these teams may (in some circumstances) actually be their competitors. Commercial organisations, in particular, were much more circumspect about sharing information, and tended to restrict this to internal sharing only. Aware of the greater openness within academia, they often showed reluctance to share information, rarely releasing it unless in response to a direct request, 'blinding' datasets and in some cases even generating dummy datasets to replace real, confidential, ones. Participants also observed differences in communications habits and behaviours, and the adoption of new technologies, between the academic and non-academic sectors. Academics were perceived by some of their commercial partners as having a different communications culture from the commercial world: one which was slower and more reliant on old technologies such as telephone or face-to-face conversations. In some cases, though, established norms and cultures could be very similar in the academic and non-academic organisations: the pharmaceutical companies and the academic laboratory working on the SGC project both reflected the strong conventions of this particular scientific field and used the same protocols, databases and electronic lab notebooks.

Project aims and partner cultures exert an important influence on information behaviours, but they also interact with several other characteristics of a collaborative project to further determine the way that researchers find, manage and share information. One such characteristic is the level of trust between the project partners, often based upon their previous experience of collaborating. This trust helped to mitigate some of the challenges which arose out of different expectations or experiences, and to make it easier to share information: as one researcher on the PRISMA project put it, 'it's much easier when you're trying to badger people for things if you have that level of contact [knowing each other personally]'. A commercial partner on the UK Reactics project said 'you need to reinforce with the academic what can and can't be discussed ... as you build that trust up you come into a position where you can divulge more and more to them.' These personal relationships were usually based upon previous experience of working together, or were created through a concerted effort at the start of the project through face-to-face meetings or mutually-trusted intermediaries. Previous experience of collaboration also helped partners to make allowances for the different working cultures, as they had a better understanding of drivers within organisations that were different from their own.

Alongside these informal, trusted relationships, most partnerships had more formal agreements in place which affect information management, usually agreed at the project outset. These arrangements might include non-disclosure agreements, formalised reporting and update structures and systems for monitoring project outputs. Often, these arrangements were put into place at the instigation of project funders: the Technology Strategy Board (TSB) was particularly keen to ensure information flow is managed in an effective way. In other instances, the arrangements might stem from the project requirements: SGC had strong information governance policies because of the highly competitive nature of its research and the need to ensure all partners were treated equally with no unfair advantages. The size and geographic spread of the collaboration also affected the formality of information governance arrangements: PRISMA, for example, a multi-national research collaboration with over 30 project researchers, constructed much more formal arrangements than Locating Communications Heritage, a small project with around five researchers based in London. The burden and legal cost of creating and observing a very formal information governance structure was also a consideration for Locating Communications Heritage researchers when they decided to maintain a much more informal arrangement.

Thus, the information environment in which researchers operate is affected by the collaboration's background, aims and objectives, and organisational cultures. Researchers' information activities reflect the specific circumstances of their collaboration, as outlined here, but there are some attitudes and behaviours which seem to permeate many of the collaborations that we studied.

First among these is a general indifference to — and in some cases, scepticism about — the importance of effective information handling, particularly around data management. Formal information governance policies — where they existed — were usually quite selective in what they covered. Non-disclosure agreements, intellectual property ownership and embargo procedures for scholarly publication were considered important as disagreements in these areas could potentially derail the collaboration; poor data management and curation was not seen as a threat in the same way. Furthermore, as projects were so outcome-focused, relatively little attention was paid to the long-term reusability of the data: this was not a success factor and therefore considered unimportant. The only exception to this was the SGC, where data publication was a key project aim. There was little evidence that failure to plan for and effectively manage information had a negative impact on any of the projects, but researchers did acknowledge that they might benefit from better information management — with the usual caveat that 'we shouldn't put any additional constraints or paperwork for us'.

Researcher expertise around systems and technologies for managing information, including data, was fairly limited. There was little evidence of researchers using centralised data repositories, virtual research environment, wikis or other systems to manage their project information and communications, although again the SGC was something of an exception due to the project aim of publishing data. Project support officers acknowledged that there was a lack of capacity and expertise, particularly in smaller organisations, but it was also clear that many participants had misconceptions about the repercussions of using online tools to manage information. Researchers expressed concerns about whether repositories could be hacked into, uncertainty about whether uploading information to an online system constituted 'disclosure' from the point of view of intellectual property, and concern that the company which owned the technical system might also claim ownership over any content that the research team put into it. These concerns were mostly borne out of uncertainty rather than fact: but researchers were also worried that implementing new systems might be costly and inefficient, particularly if they were overspecified.

Another important area of commonality across the projects was a limited use of library and information services, and a general perception that the benefits of using such services would be outweighed by inconvenience. As has been the case in other research projects, many researchers were not aware that the library was responsible for providing the published information that they used, and in discussing library services they were referring more to those which are accessed within the physical library space (Meyer et al., 2011). Library services were seen as too general, designed for students rather than researchers, and inexpert in providing the commercial information necessary for many projects. Furthermore, some participants saw the laborious selection and filtration of material as an essential part of the academic's work at the start of a project, helping them to become grounded within the field, and felt that asking a librarian for more effective search strategies might interfere with this process. While there was some use of specialist libraries by researchers who had previous experience of them through other projects, this did not usually spread to other project team members.

One library service which researchers did use — although, as we have said, in many cases they were not aware that they were doing so — was online access to published scholarly information. As might be expected, levels of access varied enormously between partners, and the relationships could be very complex. In some cases, a large commercial organisation might have better access than a non-research-intensive university. For this reason, along with the habitual openness in academia and the — incorrect — perception that the collaboration's confidentiality agreements protected all information exchange within the partnership, researchers often shared the full text of scholarly articles, despite copyright restrictions. Researchers also adopted coping strategies such as taking advantage of walk-in access at information-rich institutions — this, clearly, being a library service that they did value. In other cases, it was not considered necessary for all researchers to have access to the original literature, and much of the academics' time was taken up with producing literature reviews for dissemination to the wider collaboration. In these cases, participants often highlighted a need for strategies to help filter and refine the information to make it manageable and relevant.

4. THE ROLE FOR LIBRARY AND INFORMATION SERVICES

The previous section shows that researchers in collaborations have some fairly generic problems with maintaining effective information behaviours. Concerns about data management, information overload, new technologies and systems and the relevance of library services have all been found in previous studies of researchers working solely in academia, and librarians have begun to come up with some ways of addressing these concerns. However, as we have seen, the specific circumstances of collaborations can mean these problems present themselves in a more complex way, and that any attempts to intervene will need to take account of such circumstances. The scope of this study did not include extensive work with librarians and, as such, it would be inappropriate for this paper to make specific recommendations about what librarians should and should not do in working with collaborative research projects. But it is possible to highlight some important factors which might affect the success of any such interventions.

The research highlighted the importance of trust in collaborative projects, and its central role in securing input from partners. Trust was built through previous knowledge of partners, their introduction by a trusted intermediary, and evidence that they understood the drivers, behaviours and backgrounds of the various parties working within the collaboration. Once trust was established, information flowed more freely around the collaboration and researchers found it easier to work together. There was even some evidence that trusted members of a collaboration could begin to influence information behaviours: the PRISMA project manager, for example, was rigorous about following up information policies which 'dramatically changed the way that people worked'. Informal channels, rated by many researchers as more important than formal communication techniques, were heavily dependent upon trust. Any librarian or information professional seeking to work effectively with a collaborative project will need to ensure that they become a trusted part of the team. Early engagement with the project is crucial: this could be achieved by building links with knowledge transfer teams within universities, who will help facilitate many new collaborative projects.

Collaborative research projects are also noticeably outcome-driven, with a strong focus on the agreed outputs. This is accepted by the project's academic researchers, and they recognise that more traditional academic research outputs such as publications may have to assume a smaller role, if they go against the project aims. That said, many of them recognise that alternative benefits can flow from participation in collaborative research — whether that be a better understanding of industrial processes and the real-life applicability of academic research (Intelligent Care) or by participating in a prestigious collaboration (SGC). But the agreed output remains central to research activities, and systems and procedures therefore need to support these outputs. Services offered by librarians will likely have most success where they can show that they will support the aims and objectives of the project. It is clear that there are some such areas — for example, helping researchers to find and organise published information in an efficient manner, and perhaps also working with them to manage new data in a better way to improve project efficiency. But any such interventions must be proportional: the non-academic partners in particular are very wary about over-burdening themselves with unnecessary administration. Services will need to show that they can improve project outcomes and are tailored to the project's specific needs, even when this might mean that they fall short of best possible practice. Like the academic researchers who accept that they may not be able to publish, librarians will need to be a bit pragmatic in order to get their foot in the door: once they are there, they may be able to demonstrate the value of conforming fully with best practice and standards — or they may find that the standards which apply to academic research are less important when it comes to the aims and objectives of a research collaboration.

5. CONCLUSION

The five case studies showed that collaborative research does, indeed, operate upon its own terms. While strong disciplinary conventions, such as those in the scientific field explored by the SGC, can help academics and external partners work to the same standards and processes, in other cases local systems and behaviours prevail, meaning that researchers in different organisations end up working in quite different ways. This presents a challenge for librarians and information professionals in seeking to support researchers in collaborations, as they will be working with a number of different cultures, each of which has its own aims and objectives.

The aims and objectives of the collaboration as a whole are equally important in understanding how best to support information management and effective information behaviours. Collaborative research projects are particularly focused upon specified outcomes, and researchers in academia and external organisations direct their efforts towards achieving those outcomes. Any interventions by library and information professionals will need to demonstrate that they can contribute towards such outcomes, and will not produce additional burdens which might — as the researchers perceive it, anyway — distract from them.

Many of the information behaviours observed in these five cases will be familiar to library and information professionals: lack of awareness of library services, poor data management, attachment to known systems and suspicion about the time and effort required to begin using new ones. But the research also suggested that, in order to influence behaviour, external parties need to become trusted partners within the collaboration, and that a key way to do this is to demonstrate an understanding of the unique circumstances which govern its activities — its size, geographic dispersal, aims and objectives, participants' history of working together, and the relationship between formal and informal information channels. By recognising these conditions, and tailoring a service which takes account of them, librarians may be able to provide valuable support in making research collaborations more effective. formal and informal information channels. By recognising these conditions, and tailoring a service which takes account of them, librarians may be able to provide valuable support in making research collaborations more effective.

REFERENCES

1. BIS (2009). Fifth annual survey of knowledge transfer activities in public sector research establishments: Final report. London: Department for Business, Innovation and Skills. Retrieved October 6, 2012, from http://www.bis.gov.uk/assets/biscore/corporate/ migratedd/ publications /2/2009-5th-psre-report-v2.pdf.

2. BIS (2011). Innovation and research strategy for growth. London: Department for Business, Innovation and Skills. Retrieved October 6, 2012, from http://www.bis.gov.uk/assets/biscore/innovation/docs/i/11-1387-innovation-and-research-strategyfor-growth.pdf .

3. Bulger, M. E., Meyer, E. T., de la Flor, G., Terras, M., Wyatt, S., Jirotka, M., Eccles, K., & Madsen, C. (2011). Reinventing research? Information practices in the humanities. London: Research Information Network. Retrieved October 6, 2012, from http:// www.rin.ac.uk/our-

work/using-and-accessing-information-resources/informationuse-case-studies-humanities.

4. CIBER (2010). Research support services in UK universities: What services do researchers need and use? London: Research Information Network. Retrieved October 6, 2012, from http://www.rin.ac.uk/our-work/using-and-accessing-information-resources/ research-support-services-what-services-do-resear

5. CIBER (2011). Access to scholarly content: gaps and barriers. London: Research Information Network. Retrieved October 6, 2012, from http://www.rin.ac.uk/ node/1172.

6. HEFCE (2011). Higher education – Business and community interaction survey 2009–10. London: Higher Education Funding Council for England. Retrieved October 6, 2012, from http://www.hefce.ac.uk/media/hefce1 /pubs/hefce/2011/1125/11_25.pdf.

7. Jordan, E., Hunter, A., Seale, B., Thomas, A., & Levitt, R. (2011) Information handling in collaborative research: an exploration of five case studies. London: Research Information Network. Retrieved October 6, 2012, from http://www.rin.ac.uk/system/files/ attachments/ Collaborative_research_report.pdf.

8. Lambert, R. (2003). Lambert review of business-university collaboration: Final report. London: HMSO. Retrieved October 6, 2012, from http://www.hm-treasury.gov. uk/d/lambert_review_final_450.pdf.

9. Meyer, E. T., Bulger, M., Kyriakidou-Zacharoudiou, A., Power, L., Williams, P., Venters, W., Terras, M., & Wyatt, S. (2011). Collaborative yet independent: Information

practices in the physical sciences. Institute of Physics, IOP Publishing. Retrieved October 6, 2012, from http://www.iop.org/publications/iop/2012/file_53558.pdf.

10. RIN/NESTA (2010). Open to All? Case studies of openness in research. London: Research Information Network. Retrieved October 6, 2012, from http://www.rin.ac.uk/system/files/attachments/NESTA-RIN_Open_Science_V01_0.pdf.

11. Sainsbury, D. (2007). The race to the top: A review of government's science and innovation policies. London, HMSO. Retrieved October 6, 2012, from http://www.hm-treasury.gov.uk/d/sainsbury_review051007.pdf.

12. Wilson, T. (2012). Review of business-university collaboration: The Wilson review. London: Department for Business, Innovation and Skills. Retrieved October 6, 2012, from http://webarchive.nationalarchives.gov.uk/20120305154225/http://www. wilsonreview.co.uk /wilson-review/wilson-review.pdf.

12

CHAPTER

Supporting Teaching and Research in an Online Environment: Developing the University of London Library Model

Sally Chambers

Founded in 1837, the University of London Library (ULL) is the central research library of the University of London with particular research strengths across the broader arts, humanities and social sciences. With the development of information and communications technology, it has become a hybrid library, offering access to electronic collections but also extending access to services and its physical collections through its website and online catalogue. Research libraries now face the challenge of supporting distance learning. The ULL has risen to this challenge by being a partner in the development of webenhanced distance learning by the University of London External Programme's Virtual Campus Project. To do this, the ULL has initiated the Virtual Library Service (VLS) Project. Within the research-focussed University of London, the ULL has anticipated the potential research-support needs of its students. It is also leading a separately funded Virtual Research Environment (VRE) Project on behalf of the University Libraries Committee to address these needs and inform the University of London's activities in the virtual sphere. The University of London has created the Electronic Library Projects Team (ELPT) to undertake the VLS and VRE projects and ensure that they develop along complementary lines. This paper outlines how the ULL is supporting teaching and research in an online environment through the development of its VLS and VRE models.

INFORMATION RESOURCES FOR DISTANCE LEARNERS

The initial project undertaken by the ELPT is to develop a dedicated VLS model to support the information needs of students registered with the University of London's External Programme, many of whom are studying at a considerable distance from London. Once fully developed, the VLS will offer access to a range of information and learning resources and related services as one of the principal elements in the External Programme's Virtual Campus Project.

The University of London External Programme

The University of London, established by Royal Charter in 1836, is one of the oldest universities in the United Kingdom. It is now a federation of seventeen self-governing colleges, all of which are members of the University by choice and are empowered to award the University of London degree. The University of London External Programme was established in 1858 to make University of London degrees accessible to students who were unable to come to the University to study. The degrees awarded to students on the External Programme are of the same academic standard as those awarded internally to students studying at one of the University's Colleges in London. Currently the Programme offers over 100 qualifications at undergraduate and postgraduate level to over 28,000 students in more than 180 countries. At undergraduate level over 65% of students are currently studying in the Asia Pacific region, whereas at postgraduate level, the students are more widely distributed, with 30% in the Asia Pacific, 30% in Western Europe and almost 30% from Africa and America combined. The University of London External Programme is a distinctly global programme with the majority of students studying at a considerable distance from London. Each qualification is led by a Lead College within the University, co-ordinated by the Lead Colleges Committee. Funded wholly from student fees, surpluses are reinvested in the Programme through new course developments and through major initiatives such as the Virtual Campus Project.

The Virtual Campus Project

In October 2000, the External Programme embarked on a period of significant investment to develop new and redevelop existing qualifications, the majority of which will be delivered wholly or partially online. In parallel with these academic developments, the External Programme is currently developing an online environment, the Virtual Campus. Anyone visiting the Virtual Campus from a computer anywhere in the world will be able to access a range of information, advisory, resource and support services as would be expected on a physical campus mirrored into a virtual context. The Virtual Campus Project is a collaborative initiative bringing together three central academic services of the University of London: The External Programme, The ULL and the

University of London Computer Centre. Running from October 2000 to July 2003, the aim of the project is to provide a common infrastructure to support the development, delivery and management of internet-based courses and services for the External Programme.

The Virtual Library Service

One of the central services within the Virtual Campus Project is the VLS, which is being co-ordinated by the ELPT. Running from October 2000 until July 2003, the aim of the VLS project is to: „provide access to a range of learning and information resources and services for students registered with the University of London's External Programme from wherever they have chosen to study." Primarily, the VLS will mirror the core resources and services found in a university library, however, as more courses are delivered online, the integration of library resources within Virtual Learning Environments (VLEs) will be explored. As well as providing learning and information services for the lifetime of the Virtual Campus Project, the VLS Project will develop a model for the delivery of a sustainable service. Appropriate definitions of service will be created, including recommendations of how such sustainability may be achieved.

The VLS provides a variety of online resources and services put together in liaison with each course development team participating in the Virtual Campus Project. The principal components in the VLS are:

- Online resources: electronic datasets and internet resources targeted at each course within the Virtual Campus Project. There will also be a reference collection enabling students to find rapid reference facts to support their studies.

- Enquiry service. Available by telephone, email or web form.
- Information literacy skills. Externally developed guides and tutorials are a part of the initial VLS. The VLS project will also provide quick start guides to the resources it offers and aims to develop its own tutorials tailored to suit the specific needs of the External Programme.
- Authentication. Exploring various methods of authenticating users will be of high priority to ensure a seamless experience within the Virtual Campus.
- Web-based portal with tailored entry points for each External Programme course.

IMPLEMENTATION

Online Resources: Acquisition and Integration

One of the most significant tasks in the development of the VLS has been working closely with the academic programme development teams to determine which resources will best support the delivery of their courses. Initially, during the lifetime of the project, the VLS will provide library support for approximately fifteen courses across a wide variety of subjects including business, dentistry, economics, education, finance, healthcare, history and law. Some of the resources will be specialised, such as those for dentistry, and will only be applicable for one programme. However, other resources may be applicable to several programmes, due to the overlap between the disciplines, such as in the areas of business, finance and economics. It has been necessary to maintain a map of the resource requirements across the entire project to ensure that they are most cost-effectively acquired. In addition to paid-for resources, which so far mainly focus on subject specific databases and e-journals, the academic programme development teams have been keen to provide relevant websites and subject gateways, which are being collected and organised as part of the VLS.

The increase in both electronic resources and the online course delivery is providing the opportunity for a closer relationship between library resources and academic content. Such integration increases the necessity of the library and academic staff working closely together throughout course development. In the UK, nationally funded projects such as INSPIRAL (INveStigating Portals for Information Resources And Learning)1 and ANGEL (Authenticated Networked Guided Environment for Learning)2 , are already working closely on how such integration can be best achieved.

Support Services: Quick Start Guides and the Enquiry Service

Support for the use of the resources in the VLS is essential. Many of the External Programme students are studying both on their own and at a distance and may not be familiar with online study and electronic resources. Information about the resources and services offered is sent to each student who has access to the VLS, including Quick Start Guides explaining the basic functions of each the resources. However, it is important that an opportunity to ask questions is made available in addition to the information they receive.

The VLS enquiry service was launched in September 2001 to support the resources for the Laws Programme. Enquiries are answered, within two working days of receipt, by the Electronic Library Projects Assistant with the support of an experienced team of information and subject staff. Having a main contact person for the VLS will help to reduce the isolation experienced by students who are studying at a distance, as they will regularly contact the same person. In addition, contact details and areas of responsibility for the whole ELPT, including photographs, will be added to the VLS website to further increase the personal tone of the service. Although this is a virtual library service, it is important that students can choose the way they prefer to communicate with the library; either by email, telephone, fax, letter or perhaps even face to face if possible. Furthermore, in order to provide a dedicated service, a separate telephone number has been set up for enquiries from External Programme students. In the first instance queries are only answered between 09.00 and 17.00 (Monday to Friday) UK Time. However, due to the range of time zones in which External Programme students are studying a voice mailbox has been set up so that the library can receive telephone enquiries, as with email, 24 hours a day 7 days a week. Questions received via the VLS are recorded and analysed in order to influence the future development of the enquiry service. Already, an electronic help desk has been developed which provides answers to Frequently Asked Questions (FAQs) that have been received through the VLS enquiry service. Students are advised to consult this helpdesk before they ask a question. Since its launch, usage of the enquiry service is increasing clearly demonstrating that the students need it

Authentication

Once a student has logged into the Virtual Campus she should not be aware that any other authentication is taking place; she should not have to enter her username and password again. Any further authentication required, such as for electronic library resources, will be done behind the scenes using data exported from the External Programme's student registry. In the case of the VLS, it is essential that only authorised students have access to the resources to ensure compliance with licensing agreements. Ultimately, the VLS will provide access to a comprehensive range of resources, from a wide variety of suppliers, many of which will have different methods of provided authenticated access to their resources. A service-wide authentication is therefore a necessary component in the delivery of the service.

In the UK, the problem of multiple passwords for electronic library resources has been to some extent bridged by the development of the Athens3 , an access management system for UK Higher and Further Education. Athens provides off-campus access to a range of electronic library resources with the input of a single username and password, however, at this stage, not all resource providers will allow their resources to be authenticated by Athens. In addition, it is also not possible to integrate Athens with the authentication system that is being used to provide access to the VLEs as part of the main Virtual Campus Project. In this way, currently students are not able to login to the Virtual Campus and have access to all their resources and service without having to type in additional usernames and passwords.

For on-campus behind the scenes or quiet authentication4 access to resources is often provided by Internet Protocol (IP) recognition. For this method, the library registers the IP addresses of all the library computer terminals with the resource provider. When a student tries to access the resource from a computer with a registered IP address then they get immediate access to the resources without the necessity for any further authentication. However, for a VLS, with potential 28,000 students, it would be impractical, and not secure, to register the

IP addresses of all the students. Resource providers initially have provided remote access to such resource by authenticating students by username and password, however this will not create the seamless experience we are aiming for.

The ELPT is currently investigating a variety of authentication solutions including the use of proxy servers, which work in a similar way to IP address recognition, but without the need for the registration of student IP addresses. Attention is also being paid to the development of a second generation of Athens access management system5 . As Athens is a national access management system and is often seen as a de facto authentication standard, its use as the ultimate authentication solution for the VLS, if possible, would be preferable. However, while this investigation is in progress, most of the VLS' resources will be authenticated via the current Athens with the input of a single username and password wherever possible. In this way, fairly quiet authentication can be provided.

Integrating Access to the Virtual Library Service

The aim of the VLS is to provide seamless access to electronic library resources and services as part of the online courses. The VLS is also a coherent website providing access to electronic library resources and services independently from any online courses. When designing the VLS, a flexible infrastructure was therefore needed to satisfy two seemingly incompatible criteria. The prototype VLS was launched in September 2001. At this stage, the online course developments were still in their infancy and therefore the VLEs were not available for testing. It was therefore important to keep the design of the VLS website as open as possible to allow the pages to be integrated within the various VLEs once they had been developed. In order to increase the flexibility of the website it was decided that it should be partially database driven. Constructed in Access, the database contains the URLs (Uniform Resource Locators) for all the resources provided in the VLS. The webpages are published dynamically, using a combination of SQL (Structured Query Language) queries and ASP (Active Server Pages). This solution allows the same resource to appear on different webpages from a single URL within the Access database. The resources can therefore appear on a page within the coherent VLS website, but at the same time appear on a separate resources page within an online course. By Summer 2002, several VLE prototypes will have been developed and will therefore be available to prototype the integration of the library resources. In addition, the Electronic Library Projects Assistant will carry out an evaluation of the current site in terms of standards and good practice in web design and feedback from site users. This evaluation will take into account issues such as usability, accessibility, scalability, usage patterns and site maintenance and be used to further improve the current site. Wherever possible the feedback of the users of the VLS will input into the redesign of the website, which will be launched in time for the beginning of the new academic year in September 2002.

Issues for the Future

During the initial stages of the VLS project, considerable work has been achieved, particularly on the development and realisation of the VLS model. As 2003 approaches, the majority of the courses will be registering their first cohorts and so with the resources in place the real test of VLS will begin. With the VLS in place, appropriate definitions of service will be created, including recommendations of how a sustainable service may be delivered beyond the lifetime of the project.

SUPPORTING RESEARCH ONLINE

The Virtual Research Environment Project

A complementary project undertaken by the ELPT is the Virtual Research Environment (VRE) project, supported by a grant from the Vice-Chancellor's Development Fund to the University Libraries Committee. Running from December 2001 to November 2002, the aim of the VRE project is to: „ develop and test an integrated web-based research environment to support the needs of the e-researcher." This project builds on the work that is currently done to provide online library services for the University of London External Programme's Virtual Campus Project, as outlined above.

Although considerable work is being undertaken in the field of online teaching and learning, online research support is currently relatively unexplored. As a research-led university, the University of London is well aware of this disparity and has begun to develop online support for both research elements of taught-course dissertations, in addition to support for research degrees online.

Although considerable work is being undertaken in the field of online teaching and learning, online research support is currently relatively unexplored. As a research-led university, the University of London is well aware of this disparity and has begun to develop online support for both research elements of taught-course dissertations, in addition to support for research degrees online

Technology already exists to support the taught elements of online degree courses, both at undergraduate and postgraduate level. VLEs may be used to support a range of learning contexts from conventional classroom delivery to distance and online learning. Such software has been evaluated through the Virtual Campus Project and elsewhere. However, in order to support academic research, online software (such as groupware) must be examined for its suitability in satisfying the needs of the e-researcher.

The VRE Project is grounded in a review of current research support provision within the academic community and an analysis of the needs of the research community within the University of London. Currently a user survey is being prepared with the aim of establishing which functionalities are needed in a VRE. As the strengths of the ULL are in the broader arts, humanities and social sciences, the VRE will focus on the needs of the research communities in these areas. However, the differing needs of researchers in the physical sciences will be explored in the Research Support Review.

Principle Components

The VRE will identify and integrate principle components required for effective support of online research. This will be informed by the research support review and needs analysis. However, it is anticipated that principle components of the VRE may include:

Virtual Research Library Support

An integrated resources page within the VRE could provide software to store and compile the researcher's electronic bibliography. Personalised harvesting tools which can search both the internet and across various relevant library resources (e.g. e-journals, subscription services, even the library catalogue) on a regular basis using predetermined searches could also be included. The results of which could be added to the electronic bibliography. Furthermore, an online research skills tutorial which can be applied across a number of research scenarios could also be developed and included, for which a separate bid is being prepared.

Using the research proposal as a starting point, a subject specialist librarian from within the researchers subject area could provide a specialised gateway to relevant resources available through the library. In addition research-related questions could be answered using an electronic enquiry service and online information skills training also be developed. These services demonstrate possible synergies between the VRE project and the services delivered as part of the VLS project described above.

Research-related Information

Provision of general web-based research-related reference information, including University Research Guidelines; Citation Style Manual; Research Methods; Research Funding Sources within and outside the University of London.

Online Secure Research Repository (OSRR)

A secure environment where elements of the research, (e.g. research proposal; literature review; methodology) can be shared between researcher and supervisor could be included. In this context the VRE could contribute to the electronic infrastructure of distance research degrees that are currently being discussed within the University of London.

Online Research Support Mechanisms

The system could provide mechanisms to allow tutors and students to interact in synchronous and asynchronous modes. These mechanisms could include: discussion boards supporting subject-based discussions; integrated electronic mail; video conferencing and real-time discussion. This could encourage the development of subject-specific research communities from across the University and the provision of online seminars and presentation

Tracking of Research Activity and Achievement

Online research planning tools to track research activity and achievement could be included perhaps incorporating project management software.

Research Output Repository

Research Output Repository containing digitised or born-digital research output such as research articles, book reviews, electronic theses and dissertations could form a valuable element of the VRE.

Software Evaluation

In parallel with the identification of the principal components of the VRE, the evaluation of available software to form the VRE infrastructure is taking place. Currently a development environment to assess the suitability of Open Source software packages has been set up, although this does not necessarily anticipate the use of Open Source Software in the final implementation. It is anticipated that the principal components will be developed as modules that can easily be put together and customized by researchers. In this way it will be possible to set up different instances of the VRE for different subject areas. At present the first customized VRE-application is being set up in collaboration with the International Portal for Manuscript and Book Studies within ULL, a development of the RSLP funded project; Palaeography: Developing the National Resource.

Researcher Involvement

The generic VRE model will be tested and evaluated with a VRE user group that is currently being formed from an appropriate sample of the University of London research community. This user group will have a role throughout the development of the VRE, such as in the identification of the principal components, as well as in the testing and evaluation of the final model. A Strategic Advisory Group for the VRE project has also been set up to monitor and guide the progress of the VRE Project. The membership of the group is designed to ensure close harmonisation and exchange between the Virtual Campus Project, the Society of College, National and University Libraries' (SCONUL) E-University Task Force, the Consortium of University Research Libraries' (CURL) Task Forces on Teaching and Learning and Scholarly Communications and other relevant initiatives.

CONCLUSION

The work of the ELPT will potentially influence the ULL's future support of teaching and research in an online environment. The projects outlined above provide a test-bed for the development of new services that may be integrated into ULL's central service delivery.

RELATED WEBSITES

ULL's Electronic Library Projects Team

University of London

University of London Computer Centre

University of London External Programme

University of London Library

Virtual Library Service

This paper is adapted, with additions, from a chapter in Libraries Without Walls 4: The delivery of library services to distant users. Proceedings of an international conference held on 14-17 September 2001, organized by the Centre for Research in Library and Information, UK. Sally Chambers would like to acknowledge and thank the Electronic Library Projects Team: Ulrich Tiedau and Peter Barnes for their contribution to the writing of this paper.

REFERENCES

1. <http://inspiral.cdlr.strath.ac.uk>.
2. <http://www.angel.ac.uk>.
3. <http://www.athens.ac.uk>.
4. Stephen Pinfield uses the phrase *quiet authentication* in his article: Managing
5. electronic library services: current issues in UK higher education institutions in
6. *Ariadne*, Issue 29: <http://www.ariadne.ac.uk/issue29/pinfield/>.
7. <http://www.athensams.net>.
8. <http://www.palaeography.ac.uk/>.

13

CHAPTER

Evaluation as a Tool for Developing the Quality of Academic Libraries. Case Study at a Large and Traditional Research Oriented Scandinavian University

Kaisa Sinikara

University of Helsinki, P.O. Box 33, 00014 Helsinki, Finland

ABSTRACT

The leaders of our universities will ask today if we offer those services that a top university requires and whether the proposed resources for the libraries are worth investing. Establishing the European Higher Education Area (EHEA) requires the establishing of quality systems. Quality assurance systems will be introduced throughout the entire European academic sector and will certainly have significant impact on the development of academic organizations. The collaborative effort to evaluate European universities is coordinated by ENQA, the European Association for Quality Assurance in Higher education (Standards, 2005). Most European countries have developed a national quality assurance system. The solutions of different countries can be divided into three different classes: 1) Auditing countries[1] (institutional audit, quality audit, evaluation of quality assurance systems, enhancement-led institutional review): England, Scotland, Ireland, France, Finland, 2) Accreditation[2]: Holland, Germany and 3) Combination or another system: Norway, Switzerland, Denmark, Sweden (Audits, 2006).

THE FRAMEWORK OF THE QUALITY WORK

The leaders of our universities will ask today if we offer those services that a top university requires and whether the proposed resources for the libraries are worth investing. Establishing the European Higher Education Area (EHEA) requires the establishing of quality systems. Quality assurance systems will be introduced throughout the entire European academic sector and will certainly have significant impact on the development of academic organizations. The collaborative effort to evaluate European universities is coordinated by ENQA, the European Association for Quality Assurance in Higher education (Standards, 2005). Most European countries have developed a national quality assurance system. The solutions of different countries can be divided into three different classes: 1) Auditing countries [1] (institutional audit, quality audit, evaluation of quality assurance systems, enhancement-led institutional review): England, Scotland, Ireland, France, Finland, 2) Accreditation [2]: Holland, Germany and 3) Combination or another system: Norway, Switzerland, Denmark, Sweden (Audits, 2006).

The focus and methods are different when speaking about the quality of services, quality management or the quality assurance system.

	The Quality of the Services	Quality Management	The Quality Assurance System
FOCUS	Do the library services work well? What are the customers' needs? Are the customers satisfied?	Do the libraries provide the right services? Are they updated and cost-effective?	Focus in the entire quality system of the university
METHODS	LibQual User surveys	Peer review evaluation Self-assessment Benchmarking BSC CAF [3]	Auditing Accreditation

The focus of the article is in the possibilities of the evaluation. As an example will be used the wide evaluations that have been carried out at the University of Helsinki.

The customer viewpoint is fundamental when defining quality. The traditional customers are students and researchers, i.e. the users, but the university itself and the society can also be considered customers since they are the stakeholders who provide funding and define the policy and values for the library and information services of the university. The objective of quality management is to strengthen the comprehensive quality of user and internal services as well as the organisation and cost efficiency of the operation, such as the welfare and expertise of the staff.

EVALUATION AS A TOOL OF QUALITY MANAGEMENT

Evaluation is the systematic determination of the value (evaluation) or the comparison of objectives to the measurement of performance (assessment) based on criteria set in the relation. Evaluation is a process that aims to raise development needs and proposals. Effective planning is impossible without an evaluation component, and evaluation has little utility unless its findings are integrated into the planning process.

Evaluation embraces change and encourages libraries to treat change as a positive force. By engaging in planning and research, librarians have a better idea of the future and they can meet that future with relevant, effective, and efficient services and activities. Furthermore, they can meet new challenges and innovations, and libraries will continue to play an important and positive role in the information society (Hernon &McLure, 1990, 238-239; Hernon & Altman 1996, 15-16).

The evaluation model consists of 4 parts:

1. external evaluation organisation,
2. self-assessment,
3. peer-review with its evaluation visits and
4. public evaluation report.

In the evaluation of the study and teaching of the university and the libraries peer-review evaluation has been used. To determine the best practices the evaluation of both the assessment processes and its results is required.

At the University of Helsinki, there are 11 faculties and 20 independent institutes. About 4 300 degrees are awarded every year, of which about 400 are doctoral degrees. The number of students is 38000, and the number of staff is 7600, of which 4700 are researchers and teachers.[5]

EVALUATION PROCESSES OF THE LIBRARY INSTITUTION OF THE UNIVERSITY OF HELSINKI

The university libraries in Finland are open for everyone, which is not the case in every European country. The libraries are valuable not only for education and research at the University, but also for the society.

1993 Internal evaluation

The target-oriented development process of the library institution of the University of Helsinki began in earnest after the internal evaluation in 1993. [6] The year 1993 was dramatic from the point of view of the University of Helsinki and Finnish society as a whole: the economic depression cut resources significantly, the Ministry of Education was bringing a new budgeting model into use, and the university had to analyse its operations. All the faculties and independent institutes, including the library, rushed to be examined by the evaluation team leaded by the Vice-Rector Mustajoki. The evaluation produced more than 70 development proposals.

The library institution included more than 160 separate units. The group included some rather large libraries, like the Helsinki University Library, but the majority comprised small institution libraries and faculty libraries. The internal evaluation encouraged the creation of a presentation on four information centres and a plan for staff reductions required by the reduction in resources. The premises of the libraries were mainly old and scattered. Little attention was paid to the suitability of the library premises. The Vice-Rector Arto Mustajoki who led the internal evaluation team became the chairman of the Library Committee in 1995. The first common strategy of the library institution was accepted by the University Senate in 1998. It was decided to carry out an international evaluation in 2000.

2000 International evaluation of the Libraries of the HU

A total evaluation of the library activities at the University of Helsinki took place in the year 2000. The members of the panel were University Librarian Ian R.M. Mowat (University of Edinburg, United Kingdom), Library Director Hans Geleijnse (Tilburg University, the Netherlands), University Librarian Göran Gellerstam (University of Lund, Sweden) and Library Director emerita Kyllikki Ruokonen (Helsinki School of Economics and Business, Finland). Both the International Assessment Panel and the Steering Committee of the assessment, chaired by Professor MaijaLeena Huotari, presented extensive action plans in their assessment reports (Geleijnse et al, 2000), and the follow-up on their realisation fell to the Library Committee.

The university reserved moderate resources for the evaluation and a planning officer employed for the entire year and a part-time evaluation officer. In the evaluation several methods were used:

1. statistical and other background material;
2. self-assessment of libraries;
3. user inquiries performed on different groups;
4. the panellists' visits to the libraries (1 week + 2 days).

The evaluation panel interviewed more than 80 persons (mainly as group interviews) including decision-makers, users and actors, university management, leaders of the university administration, members of the steering group, professors, students, leaders of libraries and staff.

The most important targets for development raised in the assessment were:

- clarity in defining the National Library and the University Library sectors,
- strengthening the coordination and strategic management of the University's libraries,
- creating a common personnel strategy for all the libraries,
- improving quality,
- creating centralised supporting services.

In January 2001, immediately after the assessment was over, the Library Committee approved an action plan, which was to form the basis for developing library activities. Its relevant points were discussed in the University Senate in August 2001. At that time the Senate made a decision in principle to establish the post of Director of Information and Library Services Development and agreed on a number of other definitions of strategic policies.

THE FOLLOW-UP EVALUATION IN 2004

The goal of the evaluation was to assess the process of development within the library and information services at the University of Helsinki since the evaluation of 2000. The members of the International Panel were Library Director Hans Geleijnse (Tilburg University, the Netherlands), Library Director Gunnar Sahlin (National Library of Sweden,Sweden) and University Librarian emerita Sinikka Koskiala (Helsinki University of Technology Library, Finland). The process differed from that of the previous evaluation 2000. The steering committee was led by ViceRector Hannele Niemi, chairman of the Library Committee. The fact that the chairman of the steering group and majority of the members did not belong to the library committee in 2000 was to some extent problematic.

The follow-up evaluation was carried out more lightly than was the 2000 evaluation. The panel was supported only by the evaluation officer, who only worked part-time. The panel conducted interviews, but to considerably smaller group than in 2000. The self-assessment of libraries was an important document, as were the enquiries of the deans, researchers, teachers and students. Discussions based on the evaluation report (Geleijnse, Koskiala & Sahlin, 2004) took place in the faculties and the campuses under the chairmanship of the steering committee. A wide hearing was then arranged.

Diagram evalution process

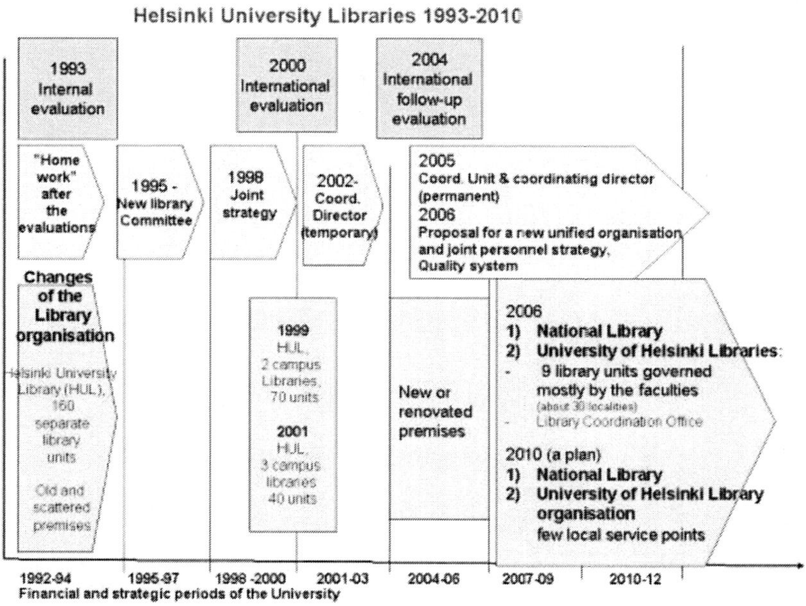

THE LIBRARY INSTITUTION IN 2006

The library institution, which functions within the sphere of the university, comprises in 2006 two large entities.

- the National Library of Finland (formerly the Helsinki University Library)
- the University of Helsinki Libraries, a coordinated network.

The University of Helsinki Libraries 2006 are located on four campuses:

- City Centre Campus (faculties of arts and humanities, behavioural sciences, law, social sciences and theology),
- Meilahti Campus (faculty of medicine)
- Kumpula Campus (faculty of science)
- Viikki Campus (faculties of agriculture and forestry, biosciences, pharmacy, veterinary medicine).
- Libraries in the four campuses

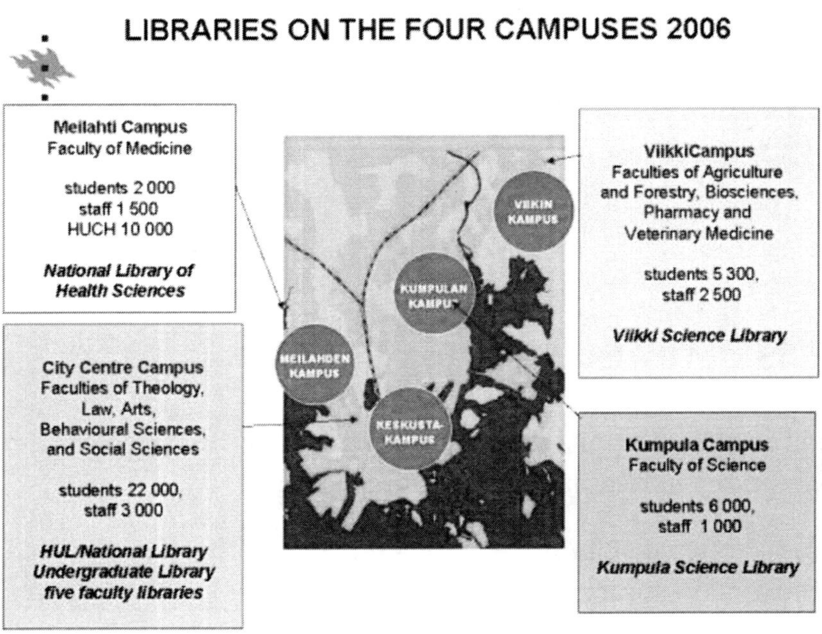

The library institution of the university has nowadays excellent, new or renovated premises. The libraries with the Library

Coordination Unit have developed a customer feedback system in a joint project (Vakkari, 2006).

'Homework' of the Library Committee and the University of Helsinki after the follow-up evaluation

The Library Committee drew up the plan of action for which the statements were requested from the faculties. The tasks of the Coordination Office were defined in the ruling of the University Administrative Office decided upon by the Senate in 2005. The Director of Information and Library Services was permanently employed and the temporary vacancies of some centralised services were permanently filled.

A lot of work has been done for the structural reorganisation of the University library staff. A plan of action to 2011 was prepared in 2006. About 25% of the personnel will retire in the coming five years. The total number of the library staff is about 500, about 50% of whom constitute the personnel of the National Library.

A big step has been taken when the reform principles of the reorganisation of the library institution with two opportunities were approached in May 2006:

- the library institution as a uniform library organisation with one managing director
- an entity with larger coordination in the administration office and with four campus libraries (one coordinating director and four campus library directors).

The model will see wide discussion during 2006. Some decisions on structural reform will be made by the Senate of the University in August-September 2006.

Difficult topics from the point of view of decision-makers

It will come as no surprise that the central problem with the faculties is how the decision-making and financing will be reorganised. As a result of historical development, these are to some extent different in each faculty. The significance and value of the library services to the humanities and to the social sciences differs from their value and significance to the sciences and medicine. This difference makes the development of the uniform library institution more difficult. For this reason, certain fears of both lines are directed at the common library institution.

The solutions depend on whether the library is considered its own special branch with well-knit connections to both research and educational functions of the university or whether it is seen only as a support service belonging to the faculties. The expansion of the electronic library has strongly changed the work of libraries even now, but it remains difficult for the faculties to accept mutual dependence on decision-making and financing.

Challenging topics from the point of view of the library directors

The biggest concern from the point of view of the library directors is connected to the freedom of action and to the independence of the library units, and especially to staff questions. Several working groups have been created for the e-library and portal development, Information Literacy, staff training, quality issues and communication. Throughout 2005, the library directors have been working with a consultant to create a common vision for the future. Important tasks have included the development of the new staff structure and the identification of common centralised tasks for the libraries.

The library coordinator's point of view

For the University of Helsinki, permanently appointing the coordinating leader to develop the library entity has been a big step. Development of the electronic library and a reasonable staff structure for the future, as well as the backup of the quality of the library services, will not succeed, if the library organisation is too scattered. At the same time the organisational model based on the coordination, is quite demanding because it includes much responsibility but little authority to carry out changes in practice. Placing the coordinator in the University Administration Office is a significant advantage because in this way, the development of the libraries can be linked to definitions of policy at the university level. On the other hand, this placing has some disadvantages, because some feel that the coordination tasks are a part of the central administration and not a part of the library services.

The financial and strategic periods of the university are quite short only three years. Every three year certain amount of the deans and committee members will change. This means possibilities but also a lot of risks for the development of the libraries.

CONCLUSION

The evaluation processes have served as a primus motor for an extensive, multidimensional and long-lasting development process in the libraries of the University of Helsinki. Without the key viewpoints on development raised in the evaluations, the same process would have been far slower to set in motion, and would have been far more difficult and burdensome to carry out than the currently ongoing process.

It can safely be stated that the library evaluations play a vital role in the development of the information services and the organisation of the University of Helsinki. On the other hand, we all know, that lot of 'homework' remains to be done after the evaluations. The process will go on.

NOTES

1. The audit is an independent external evaluation that clarifies whether the quality assurance system functions according to the objectives, and whether it is efficient and suitable for a particular purpose.
2. Accreditation (lat ad + credere): reliable and credible, presenting and confessing in relation to externally set criteria.
3. Balance Score Card – this method takes into consideration effectiveness, processes, resources and know-how, and is well-suited as an instrument for drafting a development strategy. Common Assessment Framework (CAF) provides tools for the analysis of operations (enablers) and of

results (results). See: European Institute of Public Administration (EIPA).

4. Internationally, the University of Helsinki is known for its outstanding research output for which it was selected as one of the founding member of the League of European Research Universities (LERU). Several international university rankings place the University of Helsinki among the leading universities in Europe. For example, the Times Education Supplement ranked the University of Helsinki number 18 in Europe in 2005 and the Jiao Tong ranking of the top-500 universities in the world places the University of Helsinki at number 22 in Europe.

5. The University of Helsinki: research funding 189 million euro, involved in 24 National Centres of Excellence and in 3 Nordic Centres of Excellence, 24 national postgraduate schools, and in 55 postgraduate schools organised by other universities. The share of the University of Helsinki of the universities in Finland: budget funding 23%, professors 21%, master's degrees 18%, doctoral degrees 27% and National Centres of Excellence 57%.

6. Reports: Universitas renovata 1993, Universitas renovata continuata 1994.

REFERENCES

1. Audits of Quality Assurance Systems of Finnish Higher Education Institutions, Audit Manual for 2005-2007. 2006, 37 p. (Publications of the Finnish Higher Education Evaluation Council 4). Hhttp://www.kka.fi/pdf/julkaisut/KKA_406.pdf

2. Geleijnse, Hans, Göran Gellerstam, Ian R.M. Mowat and Kyllikki Ruokonen; introduction by Aimo Virtanen: Helsinki University Libraries - Report of an Assessment panel. Helsinki : University of Helsinki, 2000. (Evaluation Projects of the University of Helsinki, 7).

3. Geleijnse, Hans, Sinikka Koskiala and Gunnar Sahlin: Follow-up Evaluation of the Library and Information Services 2004. Helsinki, 2004. (Evalution Projects of the University of Helsinki, 8).

4. Hernon, Peter and Charles R McLure: Evaluation and Library Decision Making. Norwood, N.J. : Ablex Pub.. 1990.

5. Hernon, Peter and Ellen Altman: Service Quality in Academic Libraries. Norwood, N.J. : Ablex Pub., 1996.

6. Libraries - The Heart of the University. Development Strategy for the Libraries of the University of Helsinki 1998- 2005, Helsinki : University of Helsinki, 1998. (Reports and reviews by Committees and Working Groups of the University of Helsinki, 72)

7. Mäkinen, Riitta and Aimo Virtanen: Helsingin yliopiston kirjastot - hakua vai vientiä. Arviointiprojektin loppuraportti. Helsinki : University of Helsinki, 2000. (Helsingin yliopiston arviointiraportteja, 8).

8. Standards and Guidelines for Quality Assurance in the European Higher Education Area. Helsinki : European Association for Quality Assurance in Higher Education, 2005. Hhttp://www.enqa.eu/files/ENQA%20Bergen %20Report.pdf

9. Universitas Renovata. Helsinki : University of Helsinki, 1993. 162 p. (Helsingin yliopiston arviointiryhmän raportti ja ehdotukset).

10. Universitas Renovata Continuata. Helsinki : University of Helsinki, 1994. 93 p. (Helsingin yliopiston arviointiryhmän raportti).

11. Vakkari, Mikael: Ei pidä syyttää peiliä, jos kuva ei miellytä - Kirjastojen käyttötutkimushanke. Helsinki : University of Helsinki, 2006. 70 p. (Helsingin yliopiston kirjastojen yhteishanke 2004–2006). Hhttp://www.helsinki.fi/julkaisut/aineisto/hallinnon_julkaisuja_28_2006.pdfH

Index

A

assessment, 3, 26, 27, 33, 34, 184, 186, 187, 199, 252, 253, 254, 255, 256

B

Benchmarking, 59, 95, 186, 188

biological, chemical, physical, 35

Boundary-Spanning, 86

C

cellulose oxidation, 36

collaboration, 14, 16, 22, 47, 217, 220, 221, 223, 224, 225, 226, 228, 229, 245

Convergence, 88, 101

Coordination, 91, 259

D

digital library evaluation models, 189

E

Emerging roles, 46

H

Hyperspectral imaging, 38

I

information handling, 45, 211, 214, 215, 217, 221

information society transitions, 162, 181

K

knowledge economy, 211

Knowledge organisation, 49

L

Libraries Economics, 106

library performance evaluation, 161, 162, 168, 169, 170, 172, 174, 176, 178, 180 library study, 4, 11, 16, 18, 20, 21

M

management structures, 64
meta-models, 162

O

organization design, 64, 68

P

Performed daily, 15
Positioning, 89

R

Research Data Centres, 106
Research Data Management, 84, 87, 106, 128
research libraries, 1, 2, 3, 4, 5, 102, 110, 123, 133, 138, 187, 203, 208
research support, 64

S

Specialization, 90
staff workshops, 20
Study visits, 58
Subject liaison, 73, 81

U

university libraries, 7, 64, 68, 158, 204, 208, 253
university research, 27, 133, 135, 145, 208